Learning to Pass

CLAiT Plus

2006

Jackie Sherman

Using Office 2007

www.pearsonschoolsandfe.co.uk

✓ Free online support
✓ Useful weblinks
✓ 24 hour online ordering

0845 630 44 44

Heinemann

Part of Pearson

Heinemann is an imprint of Pearson Education Limited, a company incorporated in England and Wales, having its registered office at Edinburgh Gate, Harlow, Essex, CM20 2JE. Registered company number: 872828

www.pearsonschoolsandfecolleges.co.uk

Heinemann is a registered trademark of Pearson Education Limited

First published 2010

14 13 12 11 10
10 9 8 7 6 5 4 3 2 1

British Library Cataloguing in Publication Data
A catalogue record for this book is available from the British Library.

ISBN 978 0 435 57952 4

Typeset by Tek-Art, Crawley Down, West Sussex
Cover design by Wooden Ark Studios
Cover photo/illustration © Getty Images / Lori Adamski Peek
Printed in the UK by Scotprint

Acknowledgements
The author and publisher would like to thank the following individuals and organisations for permission to reproduce screenshots in this book:

Microsoft product screenshots reprinted with permission from Microsoft Corporation.

Every effort has been made to contact copyright holders of material reproduced in this book. Any omissions will be rectified in subsequent printings if notice is given to the publishers.

Contents

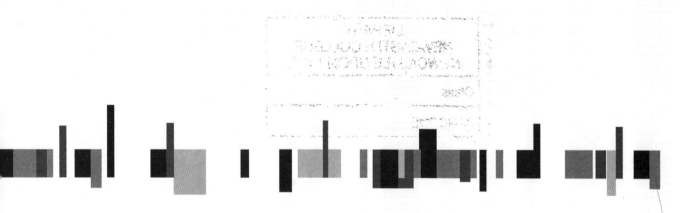

Introduction

The qualification

CLAiT 2006 is a suite of qualifications offered by OCR aimed at improving skills and confidence in general areas of Information Technology (IT) such as word processing, email, the Internet and databases. The qualification can be taken at one of three levels depending on your abilities and experience. These are:

Level 1 – New CLAiT: aimed at boosting the confidence of beginners

Level 2 – CLAiT Plus (the subject of this book): for those with some experience

Level 3 – CLAiT Advanced: for confident and productive IT users

According to OCR, 'CLAiT Plus has been designed to accredit a broad range of IT skills, ensuring that regardless of the optional units chosen, the learner will be prepared for the needs of the IT-based workplace in a comprehensive manner'.

To achieve the Level 2 certificate, you must complete assessments in the mandatory or core unit together with two others chosen from a range of options. For a Level 2 diploma, you must complete assessments in the core unit plus four others. Study normally takes place at a training centre or college, and depending on the units chosen, assessment may be online or centre-assessed. It is expected that the assignments will each take about three hours to complete.

Full details of the qualifications and accredited study centres can be found at www.ocr.org.uk .

Appropriate software

In 2007 Microsoft released a completely new version of their Office suite of programs that includes Word, Excel and PowerPoint. Until then, new versions of the programs did not differ fundamentally from the previous ones. However, with Office 2007 there has been such a complete overhaul of the look and workings of the software that it is hard for people accustomed to Office 2000, XP or 2003 to use it without difficulty. For this reason, although you can gain the CLAiT qualifications using any appropriate software, we have taken the opportunity to introduce readers to Office 2007, as well as the latest version of Internet Explorer 7.

This publication therefore covers the following CLAiT Plus units:

Core Unit:

Unit 001 Integrated e-document production

Optional Units:

Unit 002 Manipulating spreadsheets and graphs

Unit 003 Creating and using a database

Unit 005 Designing an e-presentation

Unit 008 Electronic communication

Using this book

For each unit, you will find full coverage of the underpinning theoretical knowledge and step-by-step guidance on the practical skills needed to carry out the everyday tasks required for the certificate.

After each section, you will be offered the chance to complete a short exercise to check your understanding, show that you have gained proficiency in the required skills and help identify any gaps in your knowledge. There is also a final assignment for each unit, designed to reflect the style and coverage of questions you will face when taking the assessments.

In many of the units you will also find extra material. Although not required for the qualification, this will introduce you to related topics and can be used to expand your basic skills and competencies.

Throughout the book, instructions involving clicking on-screen, pressing keys on the keyboard or selecting menu options will be shown in **bold**. For example, when you are asked to open the File menu and select the Save option, this will be shown as: Go to **File – Save**.

Some of the exercises in this book refer to documents on the accompanying CD-ROM.

Integrated e-document production

This unit concentrates on producing business documents using a range of facilities that include merging documents with mailing lists, adding data imported from elsewhere, introducing consistency to the layout of the pages and improving the appearance of any tables.

At the end of the unit you will be able to:

- ➔ use a computer to create a variety of business documents
- ➔ enter and manipulate data from a variety of sources
- ➔ import data from databases and spreadsheets
- ➔ use mail merge facilities
- ➔ create and print integrated documents
- ➔ format pages according to a house style.

Business documents

As seen in Level 1, a wide range of documents can be created using word processing software such as Word 2007. Each type of document is distinctive in terms of its layout, paper size and formatting and it is important to create documents according to the particular requirements or 'house style' of the organisation for which you work.

Types of business document

Common types of business document include the following:

Business letters

Business letters normally begin with the sender's address, perhaps in the form of a letterhead, followed by the recipient's address and the date. They start *Dear Sir* or *Madam* (or *Dear Mrs X*) and this is usually followed by the subject of the letter. They finish with *Yours faithfully* (or *Yours sincerely* if the recipient's name is known). These letters are normally printed on A4 paper that is in portrait orientation, i.e. short sides top and bottom.

Memos

Memos are internal communications and normally have a top section showing four entries:

To – the name of the recipient

From – the name of the sender

Subject – the topic of the memo

Date

The main body of the memo begins below this header.

Forms

Forms can be created offering boxes that people complete by ticking or writing in answers.

Reports

Reports are often very long documents that start with a main title and details of the writer. To make them easy to use, and in case pages get out of order, they should have a contents list, numbered pages and possibly brief details of the subject on every page.

Newsletters

Newsletters are commonly divided into columns, like newspapers, with pictures used to break up the text. They can go over several pages and stories are usually broken up so that later sections are continued on inner pages.

Invoices

Invoices are sent to customers or clients to show details of goods or services provided together with what is owed.

Fax cover sheets

Fax cover sheets have space for the names of the sender and recipient, fax numbers, the date and the subject of the fax. It is a good idea to include the number of pages being sent, in case these become separated when the fax is received.

Advertising flyers

Advertising flyers are often smaller – perhaps printed on A5 paper. They can be in portrait or landscape orientation (with long sides top and bottom), and may include borders, differently sized and enhanced text and graphics to increase their attractiveness.

Brochures

Brochures are multipage documents, usually coloured, containing full details of the item or items for sale – for example, a brochure for the sale of a country house.

Leaflets

Leaflets are often produced to provide information about specific topics such as medical advice or surgery details.

Itineraries

Itineraries display details of journeys and may include timetables, locations, hotels and other travel arrangements.

Questionnaires and surveys

Questionnaires and surveys contain lists of questions that may have answers to write in or numerical grades to select.

Paper size

Paper sizes vary considerably and it is usually your personal choice which you use. The default setting in word processing applications is A4, but this may not be appropriate for many documents. If you are using different sized paper, make sure you have placed the correct sized paper in the printer tray and checked and changed the settings for your printer before taking a hard copy.

Common paper sizes in the UK are:

A3 = 42cm x 29.7cm – used for drawings, posters and large tables
A4 = 21cm x 29.7cm – used in offices for letters and forms
A5 = 21cm x 14.8cm – notepad size
A6 = 10.5cm x 14.8cm – postcard size

Check your understanding 1

1 Try to find examples of different business documents.

2 Compare their style, layout, paper size, margins and orientation so that you are familiar with the differences.

Creating new documents

After launching Word 2007, you are always offered a blank document. Here you will have basic settings – the defaults – such as font type, font size, margins and orientation pre-set, but these can be changed to suit your work. However, you need to start designing the document from scratch.

For many business documents, it is quicker to customise a standard layout provided by one of the many templates within Word. Normal documents using these layouts can be created and saved in this way with the underlying template remaining unchanged.

create a document based on a template

1 Open **Word**.

2 Click on the **Office** button.

3 Click on **New**.

4 In the window that opens, click on **Installed Templates** in the index on the left.

5 This displays a range of document types. Click on any one to see a preview.

6 Select your preferred template and click on **Create**.

7 When the template opens, customise it to suit your needs. In some cases, there will be boxes to click into and you can delete and retype any entries.

8 Save it as a normal Word document.

Templates created by you   Word templates

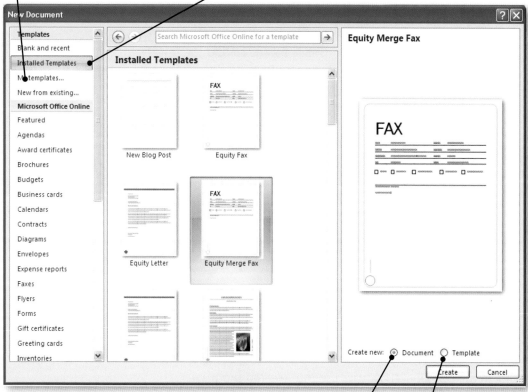

Fig. 1.1 Templates   New document   New template

Opening files created previously

There are two basic methods for opening files:

- Locate and open a named file from the desktop.
- Open a program such as Word 2007 and then use the facilities within it to locate and open the file.

Locating and opening a named file from the desktop

With the file name visible, double click and it will open into an appropriate program with which it is compatible and has been associated. For example, a **workbook (.xlsx)** and **comma delimited file (.csv)** will normally open into **Excel** and a **bitmap** image file into **Paint**.

If you want to open a file using a specific program that is not set as the default, you can create a new association.

change which program opens a file

1 Right click on the file name on the desktop.
2 Select the **Open With** option.
3 Select a program from those listed if one is appropriate.
4 To open the file with a different program, click on **Choose Program**.

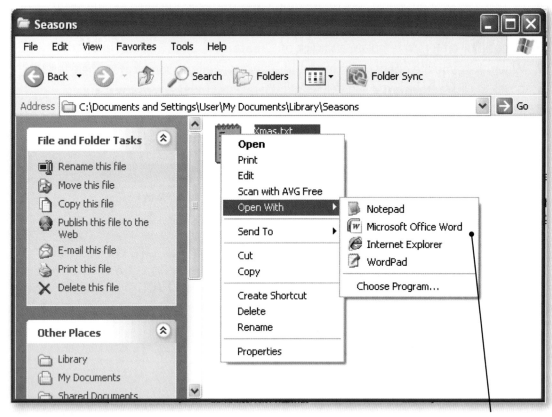

Fig. 1.2 Choose Program

Suggested programs

5 In the new window, select your preferred program from those available.
6 Click on **OK** to open the file.
7 If you want to use the selected program when opening all future files of the same type, click in the box labelled *Always use the selected program.*

Locating and opening a named file within a program

open a file within an application

1 Launch the program, for example Word 2007.

2 Click on the **Office** button.

3 Click on **Open**.

4 If the target file is listed in the Recent Documents pane, click on its name. Otherwise click on the **Open** button. (You can also hold down **Ctrl** and press **O**.)

5 When the **Open** window appears, search for the folder containing the target file.

6 For files on a different drive, such as on a CD or networked server, click on **My Computer** and then open the correct folder to search inside.

7 Double click on the file or click on the file name and then click on the **Open** button.

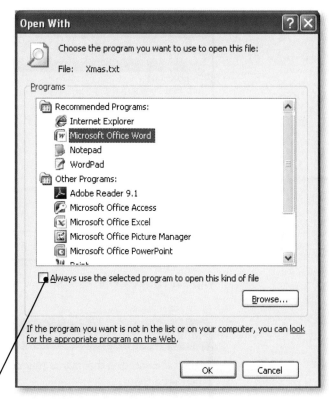

Suggested programs   **Fig. 1.3** Open with list of programs

Note that when using this method to locate different types of file, you may need to change the *Files of type* to *All files*, or select a specific file type, for example **.txt** (Plain Text File) or **.rtf** (Rich Text File). Otherwise you will only be able to search for Word 2007 documents.

Locate files on other drives

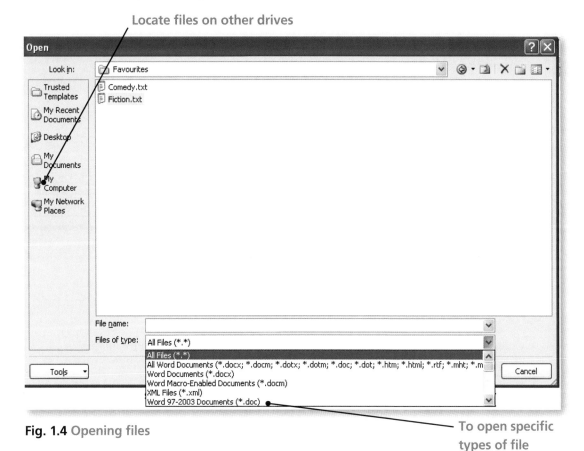

Fig. 1.4 Opening files

To open specific types of file

8 You may be offered a **Convert file** box, so select the correct file type and then click on **OK**.

Fig. 1.5 Convert File

When you use a particular software application there is a limit to the types of file you can open. If you try to open one of the wrong types – for example an image file using a word processing application – you will see an error message saying it cannot be read, or strange symbols and a box offering the chance to convert the file into one that is readable.

Some types of file are known as **generic files** as they can be read by a number of different programs. These include simple text files such as **.txt** and **.rtf** files, or data files such as **.csv** (comma delimited).

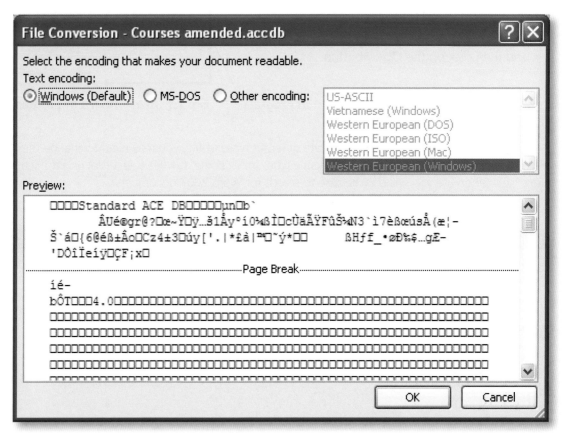

Fig. 1.6 Wrong type of file

Check your understanding 2

1 Open the image *Sunset* on the CD-ROM accompanying this book into the default image editing program on your computer and then close it.

2 Now open it using a different image editing program, for example **Paint, Paint Shop Pro** or **Windows Picture** and **Fax Viewer**.

3 Finally, open **Word 2007** and then try to use the normal open facilities to open the image file *Sunset* within this application.

Executable files

One group of files stored on your computer that are not related to specific software applications are the **executable** or **.exe files** that run the programs. When you download an application from the Internet, for example, it will be in the form of a folder of related files that include one or more main .exe files as well as images, help documents and data files that are all needed for the program to run properly.

When you double click on an .exe file on the desktop, you will start the program.

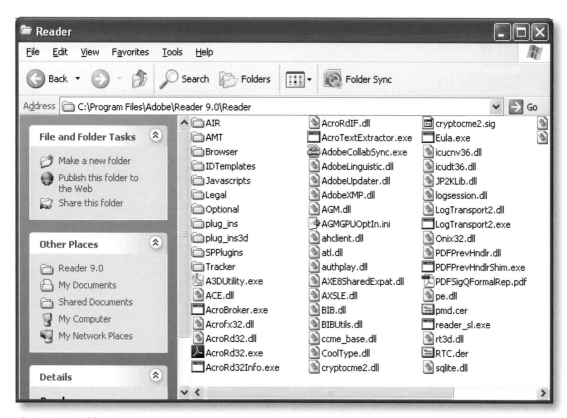

Fig.1.7 .exe files

Screen prints

To display new settings or show that folders have been created or accessed, you may need to provide evidence in the form of screen prints. Although covered at Level 1, here is a reminder of the steps to take.

take a screen print

1 Press the **Print Screen (PrtSc)** key on your keyboard for a picture of the whole screen.
Or

2 Hold **Alt** as you press the key to print only the active window.

3 Open a new document or image editing program and press **Ctrl + V** (or select **Edit – Paste**).

4 The screen image will appear.

5 Save and/or print the file in the normal way.

Note that when carrying out an assessment, it is a good idea to create and save a file with a name such as *Screen prints* and keep this minimised. Whenever you need to take a screen print, open the file and paste in the next image.

Saving

In the same way that only certain files can be read by particular programs, you can only save files in a limited range of file types. These are listed in the **Save as type:** box in the **Save As** window and may include templates, web pages and earlier versions of the software.

Note that if you save a Word document in a **Plain Text** format it will lose its formatting. Saving in a **Rich Text Format** will maintain some formatting but not all.

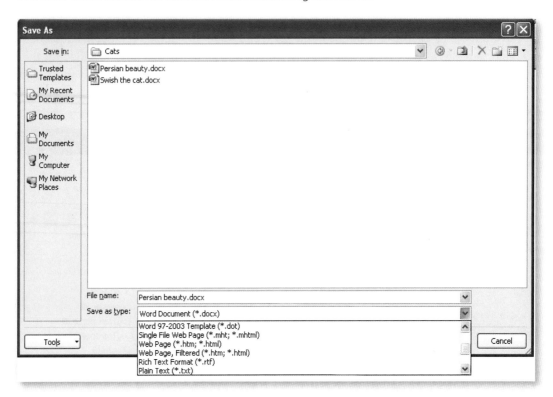

Fig. 1.8 Save as type

Printing

Before printing any documents, take the following precautions:

1 Proofread and spellcheck carefully on screen.
2 Check in print preview to make sure documents have no obvious spacing or other errors.
3 If necessary, change page orientation, margins or paper size at this point.
4 Click on **Shrink One Page** if a small amount of text has gone over onto another sheet.

Layout options

Fit text on fewer pages

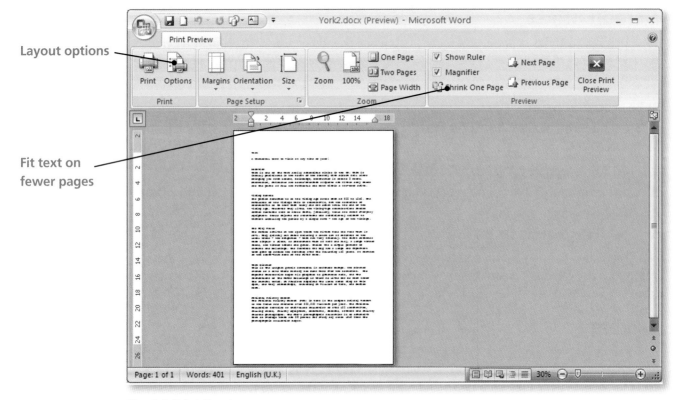

Fig. 1.9 Print Preview

5 Open the **Print** dialog box and change any settings if the number of copies, pages or appropriate printer has not been selected.

6 Check the printed output and, if necessary, correct the document on screen and print again.

use the spell checker to check a whole document

1 On the **Review** tab, click on the **Spelling and Grammar** button.

2 Make sure English (UK) is selected as the dictionary language.

3 When the first word is highlighted, click on an **Ignore** option to retain it in the document, or change it as follows:

   a Click on a suggested alternative in the **Suggestions** window

     or

   b Click into the red highlighted word and correct it manually.

4 Click on a **Change** button to update the document and replace the word once (**Change**) or wherever it appears (**Change All**).

5 For a word you will use repeatedly, click on the **Add to Dictionary** button so that it is accepted in future.

6 When the entire document has been checked, click on **Close** to close the window.

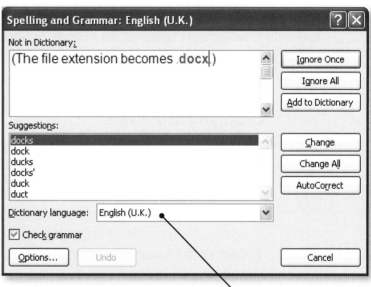

Fig. 1.10 Spell check   Correct spelling

Adding to the dictionary

If you want to add a word to the dictionary so that it is recognised and not flagged up in future, you can also add it manually via the **Proofing** option.

add a word

1 Go to **Office – Word Options – Proofing**.

2 Click on **Custom Dictionaries**.

3 Click on **Edit Word List**.

4 Type the word in the box and click on **OK**.

5 To remove a word, select it in the list and click on **Delete**.

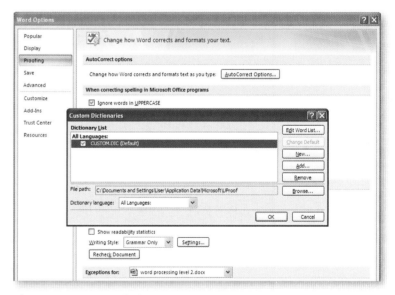

Fig. 1.11 Custom Dictionary

Check your understanding 3

1 Open **Word 2007**.

2 Open the text file *York* on the CD-ROM accompanying this book.

3 Save it as *York2* in a Word 2007 document format. (The file extension becomes .docx.)

4 Check the document carefully and print one copy.

Minimising saving

Each time you save a file, it takes up space on your hard disk or storage media. It is therefore good practice to limit saves. In particular, if files contain imported data, you may only need to save the original documents and not the finished, merged one as well. Instead, you can link the two so that, when one opens, the imported data from the other document will be displayed and can even be updated automatically.

Archiving

Archiving is the general term used for keeping copies of important files. You can do this in a number of ways:

- store copies separately, for example copy them to a new folder, disk or drive
- change the file attributes
- compress or zip files.

Changing attributes

The archive attribute is recognised by the operating system when you try to make copies. If a file is copied successfully using a backup utility, the system will clear the attribute from the original file. If the archive attribute is still turned on, it means a copy has not yet been saved in this way.

set archive attributes

1 Right click on the file on the desktop and select **Properties**.
2 Click on the **Advanced** button on the **General** tab.
3 In the **Advanced Attributes** window, make sure there is a tick in the **Archive** checkbox.

Fig. 1.12 Archive Attributes

Compressing files

One way to save space is to store copies of important files in a compressed or zipped format as an archive file. The archive can then be saved on removable storage media or sent attached to an email. When an archive is received or when you want to access the contents, the compressed files can be extracted and then treated in the normal way.

Windows XP machines contain the software to archive files directly. For earlier operating systems, you will need to use a program such as **WinZip** that is available free on the Internet.

zip files using Windows XP

1 On the desktop, select all the files you want to compress. Hold **Ctrl** as you select individual files if they are not adjacent.

2 Right click on any selected file and select **Send To – Compress (zipped) Folder**.

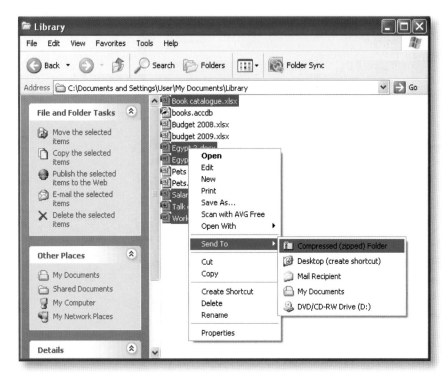

Fig.1.13 zip files

3 A new yellow folder containing the compressed files will appear. The icon has a zip across it, and it will be labelled with the same name as one of the selected files. The file extension will be **.zip**.

4 Rename the archive to avoid confusion.

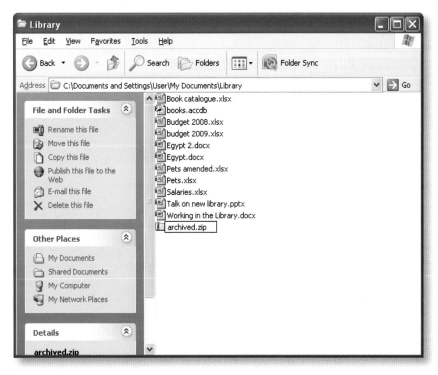

Fig. 1.14 Rename archive

5 If you double click on the archive, you will display copies of all the selected files.

6 You will be able to see that the whole archive takes up less space than the individual files would normally.

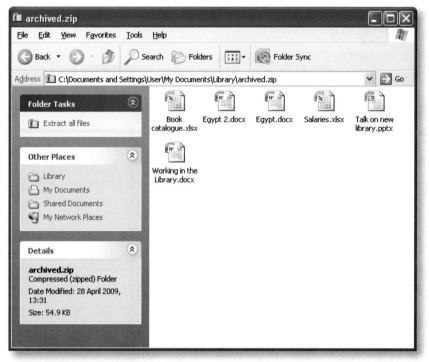

Fig. 1.15 Opened archive

work with archived files

1 Open the **archive**.

2 Double click on an individual file to open it as normal.

3 To save all the files outside the archive, click on the *Extract all files* link in the **Folder Tasks** pane. This starts the **Extraction Wizard**.

4 Click on **Browse** to select a destination for the files. At this stage, you could create a new folder to contain the extracted files. Right click on it when it appears in the folder list to give it a suitable name.

Fig.1.16 Extracting

5 Click on **Next** and watch the progress of the files being extracted.

6 Click on **Finish** and choose whether to view the extracted files or not.

Check your understanding 4

1 Select any three files saved in **My Documents**.

2 Archive them and name the new archive *Compressed*.

3 Open the archive to check its contents.

4 Extract the contents to a new folder named *My zipped files*.

5 Open this folder and take a screen print of the contents.

Fig. 1.17 Archive in my documents – Step 2

Fig. 1.18 My zipped files – Step 5

File Protection

With important documents, you can protect them from accidental alteration or deliberate editing in two main ways:

● Make them **read-only**. In this mode, they can only be viewed.

● Use **password protection**. Without the password no one will be able to open the file.

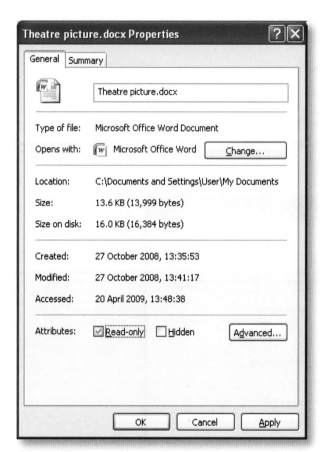

Fig. **1.19** Read-only

make a file read-only

1 Right click on the closed file on the desktop.
2 Select **Properties**.
3 Click on the **Read-only** checkbox.
4 Click on **OK**.
5 To reverse the effect, open the **Properties** box and remove the tick.

Fig. **1.20** Encrypt Document

add password protection

1 Click on the **Office** button.
2 Select **Prepare**.
3 Click on **Encrypt Document**.
4 In the dialog box that opens, type a password, making sure you choose something you will remember in future.

5 Click on **OK**.
6 In the **Confirm Password** dialog box, re-enter the password and then click on **OK**.
7 To save the password protection, you must save the file.
8 To remove the password at a future date, open the **Encrypt Document** dialog box, delete the password showing as dots and then click on **OK**.

Fig. **1.21** Opening password protected documents

open a document that is password protected

1 Start to open the file in the normal way.
2 When the dialog box appears, type the password in the box provided.
3 Click on **OK** and the document will open.

Check your understanding 5

1 Start a new document.

2 Save it as **Protected**.

3 Set the attributes as Read-only.

4 Take a screen print showing that this setting has been applied.

5 Protect the file **Protected** with the password **remember**.

6 Save and close the file.

7 Open **Protected** and remove the password protection.

8 Save and close the file.

Mail merge

Instead of sending personalised letters or invoices individually to a large number of people on a mailing list, you can use Word's **mail merge** facilities. This allows you to create a single main document and then merge this with the details of everyone on the mailing list. The main document is the general term used for any type of document such as a letter, invoice or mail shot. You can use the same facilities to create labels or envelopes.

Data from the mailing list will be inserted in the form of **fields** – in other words the headings under which the data is stored: Surname, Town, Postcode, etc.

The completed main document might start like this:

<First Name> <Surname>

<Address Line 1>

<Town>

<Postcode>

Dear **<First Name>**

When merged, each letter will include details within the specified fields drawn from a separate record in the database.

LETTER 1	LETTER 2
John White **3 The Willows** **Manchester** **M28 4LP**	**Mary Sheldon** **26 Ash Road** **Worthing** **BN13 4EP**
Dear **John**	Dear **Mary**

As well as names and addresses, the data source can include other information about the individuals if this will be needed in the final, merged documents – for example the library books they have borrowed, their salaries, dates they are starting a course, and birthdays.

carry out a mail merge

1 Create or open the main document.

2 Create or locate the data file containing all the names, addresses and other personal information. This can be in the form of a word processed table, database file or spreadsheet, as long as it is set out appropriately. This means that it must have column headings (**field names**) in the first row and one data entry for each field.

3 Merge the documents.

4 Print your merged documents directly, or save them in the form of a new, single merged document known as a **form letter**. This will have all the letters set out on a new page with each letter displaying one person's details. These merged letters can be stored separately from the source data file.

Creating the main document

If you want to use a document that has already been created it is important to remove all personal details such as names and addresses as these will be drawn from the data file. Alternatively you can start by simply opening Word and creating the document directly.

create a main document

1 Open Word and go to **Mailings – Start Mail Merge** on the ribbon.

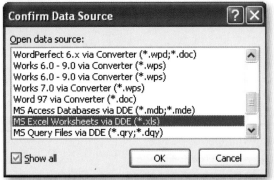

2 Select **Step by Step Mail Merge Wizard** to open a help pane alongside the screen.

3 Step 1 – select **Letters** for any type of document other than labels or envelopes and then click on **Next**.

4 Step 2 – click on *Use the current document* if no letter has been created.

5 Click on *Start from existing document* to open a saved document on screen.

6 At this stage, you could type any details that will be included in all the merged documents, or leave this until later.

Fig. 1.22 Mail Merge – Step 2

Accessing the data source

Word 2007 offers the option to create a special Office database file during a mail merge. However, it is common in many organisations to use data already stored on a large database elsewhere. You can therefore choose whether to create the data source file from scratch or link your main document to an existing one.

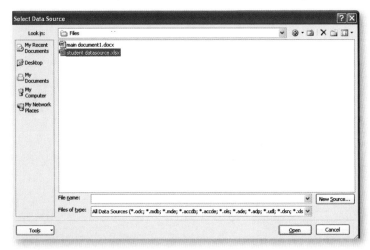

Fig. 1.23 Browse for data source

access an existing data source

1 Click on **Next**.

2 Step 3 – to link to an existing database, select this option and browse for the file. As you may need to look for a non-Word file such as an Excel spreadsheet or Access database, the system is set to search for 'all data sources' automatically.

3 When located, select the file in the main window and click on **Open**.

4 You may be asked to confirm certain details such as the specific sheet of an Excel Workbook on which the data is held.

5 When you click on **OK** you will return to the main document and the name of your data source file will be displayed in the wizard pane.

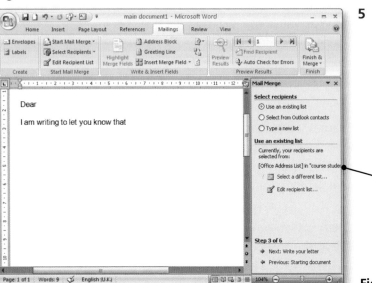

Fig. 1.24 Data source file in wizard

Check your understanding 6

1 You are going to carry out a mail merge. Use the document *Lunchtime* as the main document.

2 Select the file *Lunchtimedata* as the data source.

create a new data source

1 At Step 3 click on **Type a new list – Create**.

2 A **New Address List** window will open displaying a range of headings.

3 Type the details of the first person on your mailing list in the appropriate boxes in the first row and then click on **New Entry** to start typing in record 2.

4 Continue entering records until all the details have been added.

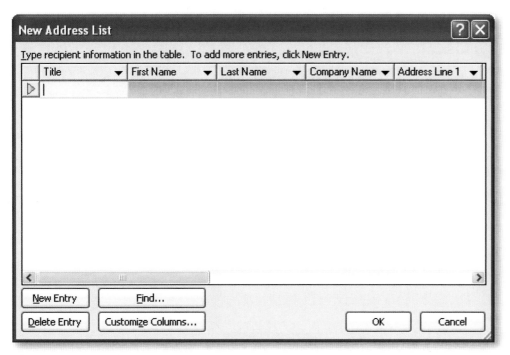

Fig. 1.25 New Address List

5 To create your own field/column headings, click on **Customize Columns**. You will be presented with a list of all the fields.

   a Click on **Delete** to remove unwanted fields from your data source file.

   b Select a field and click on **Add** to add a new one *below* it in the list.

   c Click on **Rename** to amend the field name.

   d Click on **OK** when the list is complete.

6 When all records have been entered, save the data source with an appropriate name and in a location that is easy to find. (Note that by default the file will be saved into the **My Data Sources** folder in My Documents.)

7 Back in the main document, the name of your new data source file will be displayed in the wizard pane.

8 At any stage, as long as you are not engaged in a mail merge, you can open the data source file to add or amend records.

Fig. 1.26 Customize headings

Merging data

Once you have linked your main document with a data source, you can start merging the two files. This involves inserting the correct field (sometimes referred to as a merge code) from the data source into the document at an appropriate place.

insert fields

1 Click on **Next** to start writing the letter.
2 Step 4 – you can now start or complete the document by inserting fields as appropriate.
3 Click in the place in the letter where you want to add details drawn from the data source file and then click on **More items**.
4 Select the field and click on **Insert** to add it to the document.
5 Click on **Close** to close the **Insert Merge Field** window and continue writing the letter.

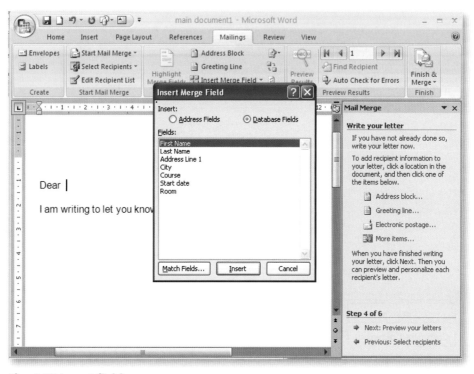

Fig. 1.27 Insert fields

6 Where relevant, use the shortcuts offered:
  a Click on **Address block** to add the various address details in your preferred format.
  b Click on **Greeting line** for a *Dear...* entry.

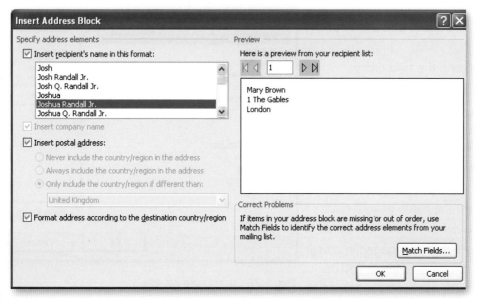

Fig. 1.28 Address block

7 When complete, your document will contain a mix of field names and normal text. The fields will be within << >> chevrons.

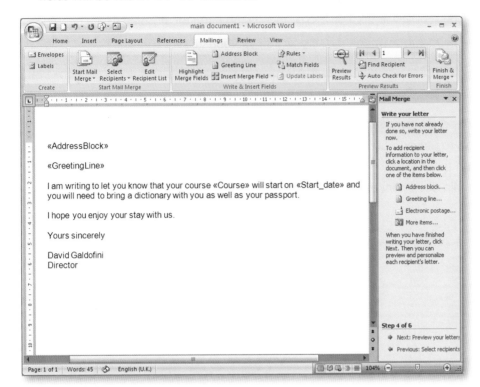

Fig. 1.29 Finished document with fields

Check your understanding 7

1 You are going to complete setting up the *Lunchtime* main document so, if you have not yet done so, first carry out 'Check your understanding 6' to link the document to the *Lunchtimedata* source file.

2 Insert all the fields in the appropriate places within the memo.

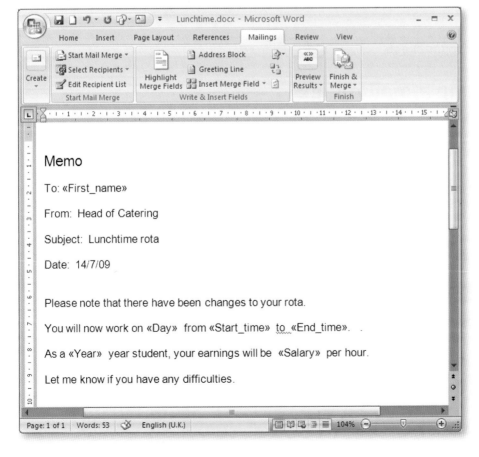

Fig. 1.30 Lunchtime merge1

Reviewing your documents

To see what the letters will look like when printed, you need to move on to the next step and preview them. If you spot any errors, you can correct them before the documents are printed.

At this stage, use normal formatting tools to select an appropriate font type and size, set the margins and page orientation and make sure text does not stretch onto a new page. One common mistake is to insert fields without the right gaps between them. This usually only becomes clear when you preview the letters with actual names and addresses inserted.

preview merged documents

1 Click on **Next** to move to Step 5.

2 Move through the records by clicking on the **Next** and **Previous** buttons.

3 Format the main document and correct any errors.

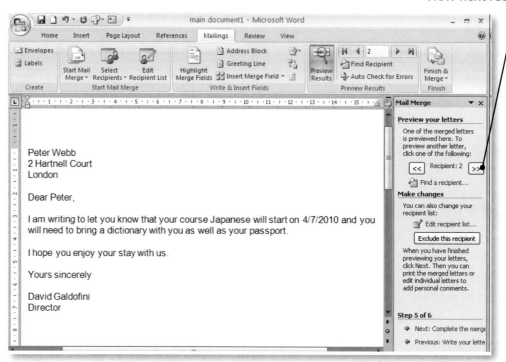

Fig. 1.31 Preview documents

Completing a mail merge

Once the letters have been completed, you have three choices:
- Print them directly.
- Save them as a single merged document temporarily titled *Letter1*, with each individual letter displayed on a separate page. As this duplicates the data, this alternative should normally be chosen only when you need to access or print the merged documentation separately from the data source.
- Save and close the main document and data source files without printing. When you open the main document later, it should be linked to the data source file. To continue with the mail merge, choose the **Yes** option.

Fig. 1.32 Opening main document after closing

print merged letters

1 Click on **Next**.
2 Step 6 – click on the **Print** option and decide whether to print all or selected letters or just the current document open on screen.
3 The **Print** box will open and you can change settings and print as normal.
4 Save the main document if you want to access it again in the future.

create a new document

1 At Step 6, select the *Edit individual letters* option.
2 Include all or some of the records and produce a new, merged document.
3 To make the merged document easy to find in future, save it with a name other than the default.

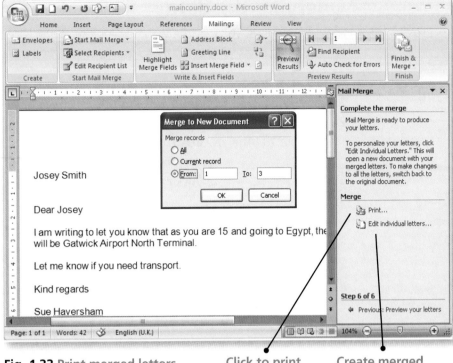

Fig. 1.33 Print merged letters

Click to print

Create merged document

Check your understanding 8

1 Following on from 'Check your understanding 7', preview the memos and correct any spacing or other errors.
2 Open the memo addressed to Nalia on screen.
3 Print a copy of this current document only.

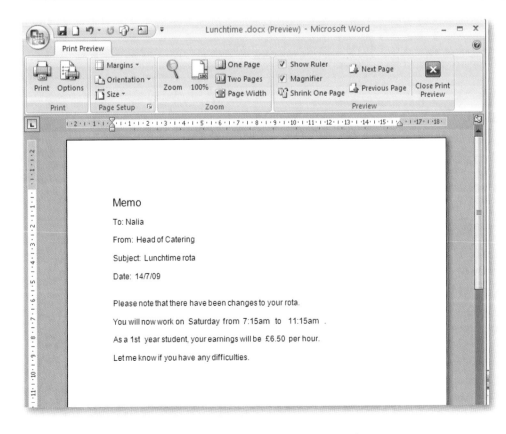

Fig. 1.34 Printed merged letter

Querying the data

For some mailings, you may not want to send out all the letters to everyone in the database. You can therefore select recipients before printing or merging your documents.

query the data

1 Click on the **Edit Recipient List** button.
2 Click on one of the drop-down arrows in the field name box to select a single entry, for example 18 (Age) or Greece (Destination).
    or
3 Click on **Filter** to open the **Filter and Sort** window. You can now complete the following:
    a **Field** – select the appropriate category, for example *Age*.
    b **Comparison** – choose from the options such as *less than, equal to,* and *more than.*
    c **Compare to** – enter the criterion on which to filter, for example *less than 18.*
4 Click on **OK** and continue with the mail merge. The records will include only those meeting your chosen criteria.

Edit recipients

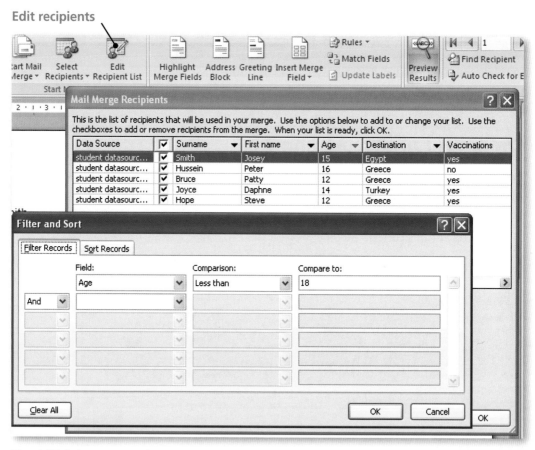

Fig. 1.35 Select merge data on 1 criterion

Using the mail merge ribbon

When you click on the **Mailings** tab, you will see a range of mail merge buttons along the top of the screen. You can use these to carry out a mail merge directly, without working through the wizard.

1 Click on the **Start** button to select the type of main document.
2 Click on **Select Recipients** to browse for the data source file or create a new one.
3 Filter out particular records by clicking on the **Edit Recipient List** button.
4 Insert fields using the **Insert Merge Field** button, or add a full address or greetings entry.
5 Click on **Preview Results** to see the merged letters, and click on the arrows to move backwards and forwards through the records.
6 Click on **Finish** to print or merge the documents.

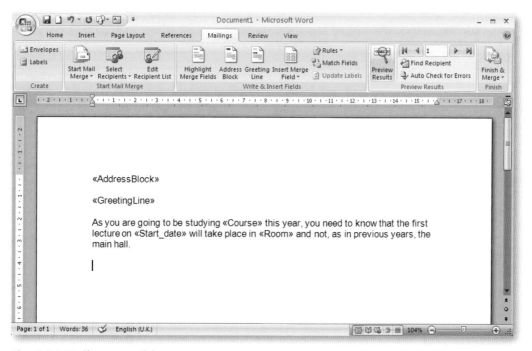

Fig. 1.36 Mail merge ribbon

Check your understanding 9

1 Start a mail merge using a new blank document.

2 You are going to create a new recipient list. Remove all unwanted fields, add the four listed below and then enter the following six records:

Florist	Date	Town	Number
Floribunda	September	London	250
Flowers for all	March	Manchester	180
West London Flowers	August	London	310
Floribunda	August	Exeter	156
Say it with	October	Norwich	320
Hazels	June	London	455

3 Save the data source file as *Plantdata*.

4 Create the following main document and insert the fields where shown:

Fig. 1.37
Example of florist main document

5 Print only the merged documents for orders going to London (there should be three).

6 Save the main document as *Plant mailmerge*.

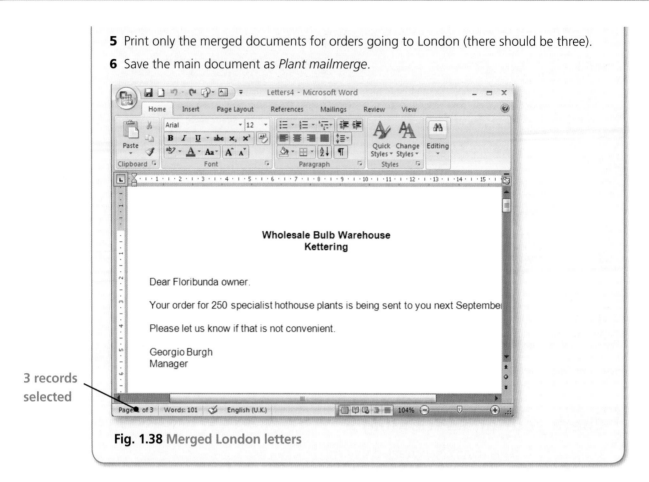

3 records
selected

Fig. 1.38 Merged London letters

Integrating documents

It is often necessary to import data into business documents that has been created and saved already. Word enables you to bring in text, images, graphs and other data very easily.

The simplest method is to use **Copy** and **Paste**, but in some cases it is better to use the insert function. Insert options vary, depending on the source of the data you are introducing, and you must take care that the imported item is displayed correctly and does not, for example, extend into the margins or get split across pages.

Inserting text files

When you import text documents, the text will retain its original formatting and you can either leave this or apply the main document formatting as appropriate. If the imported file contains any objects such as images, these will also appear.

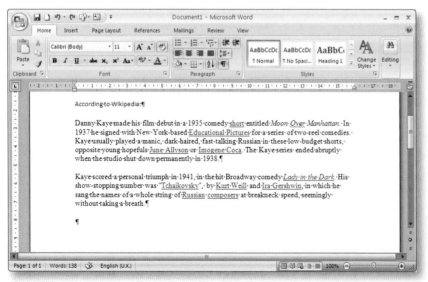

Fig. 1.39 Inserted text file

When importing a text file at the top of a page, there should be no paragraph markers visible when you click on the **Show/Hide** button. For an insertion between paragraphs, there should be paragraph marks before and after the inserted text.

insert a text file

1 Click where you want the text to appear.

2 Click the **Insert** tab.

3 Click the drop-down arrow next to **Object** in the Text group on the ribbon.

4 Select **Text from File**.

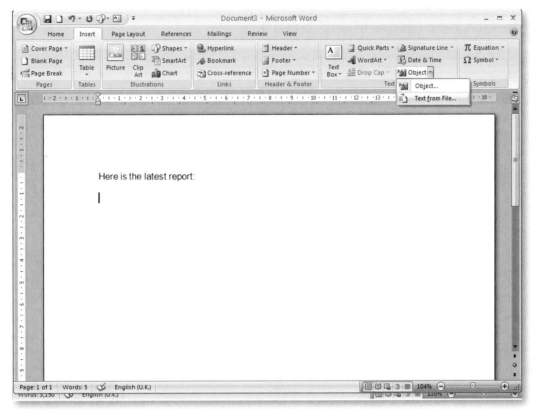

Fig. 1.40 Insert text file

5 In the **Insert File** window that opens, navigate to the file you want to insert. When looking for files other than Word documents, for example **rich (*.rtf)** or **plain text (*.txt)**, first change the *Files of type* to *All Files* or select the relevant type.

6 Select the file name in the window and click on **Insert**.

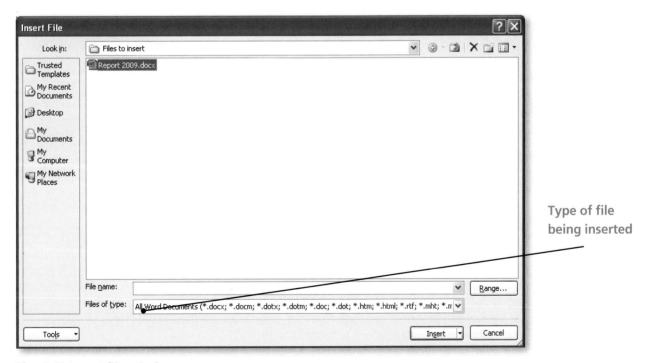

Fig. 1.41 Insert file window

7 You will return to your main document and the new text will have been added.

8 If necessary, correct the text spacing. It may help to turn on the **Show/Hide** function that reveals all hidden formatting symbols.

Check your understanding 10

1 Open the file *Sportsday* on the CD-ROM accompanying this book.

2 Click on the line below the words *who won a prize* and insert the contents of the file *Results*.

3 Enter the name *Charlie Foxton* at the end of the document.

4 Save as *Integrated*.

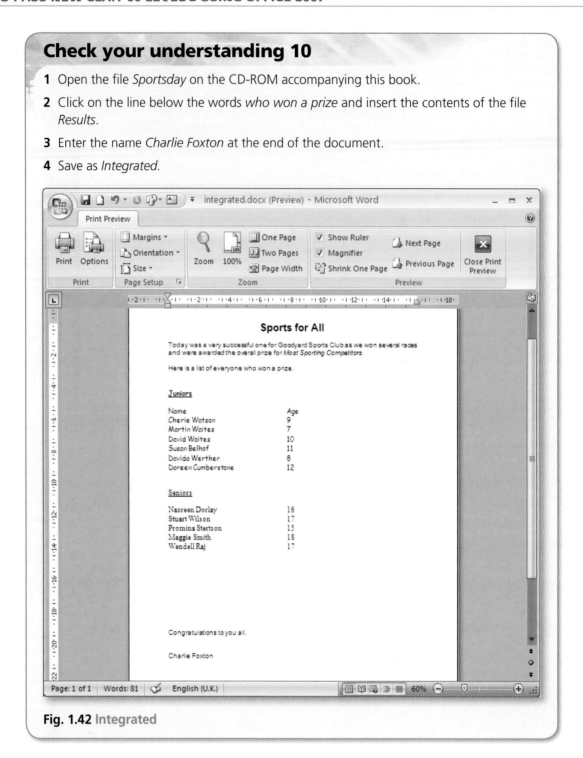

Fig. 1.42 Integrated

Importing data files

Information that you want to import into a Word document may be held in something like a spreadsheet or .csv file. When imported directly into a Word 2007 document, it will normally appear as a table, but the editing options are different from normal Word tables. This is because Word uses **OLE (Object Linking and Embedding)** technology so that, when double clicking, you will be able to use formatting tools from the original application. You will also be able to format the object to a certain extent with Word tools.

Note that exactly the same process is involved if data is stored in a PowerPoint file or if you want to import a chart.

Word 2007 converts most imported data into a table, but if it appears as text, you can use the facilities to convert it into a normal Word table. For example, a .csv file opened with Excel appears as a spreadsheet but in Word can display text entries separated by commas.

Fig. 1.43 CSV and Excel

There are two ways data files can be imported:

- **Linked** – if the original data is changed, you will be able to update the data imported into your main document as it is stored in the source file.
- **Embedded** – here you can only change the data in the main document manually as the data has become part of the Word file.

insert a data file

1 Click where you want the data to appear and go to **Insert – Object**.

2 Click on the drop-down arrow and select **Object**.

3 Click on the tab labelled **Create from File** and then click on **Browse** to navigate to the data file.

4 Click on **Insert** and its name will appear in the **File name** box.

5 Click in the **Link to file** checkbox if you want to keep the imported data up to date. Otherwise it will be embedded.

6 Click on **OK** and the data will appear.

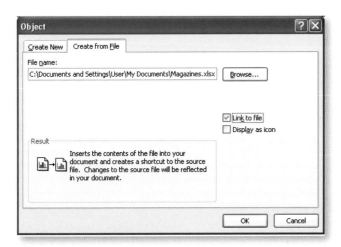

Fig. 1.44 Insert data file

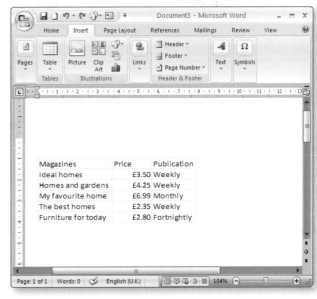

Fig. 1.45 Imported data file

edit imported data

1 Double click on the table to edit it.

2 For an embedded (non-linked) table, you will display all the tools to allow you to format the numerical data, insert or delete cells and perform calculations.

3 Scroll bars will enable you to move round the sheet and you can click on a sheet tab to change to a different sheet if the data is elsewhere in the workbook.

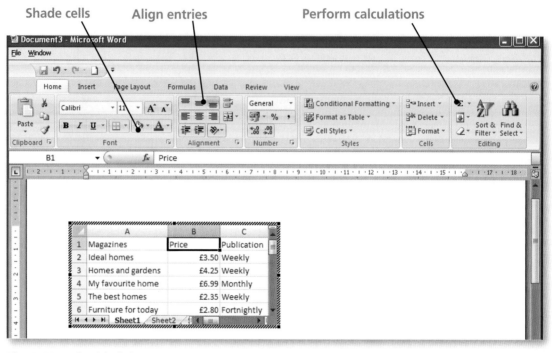

Fig. 1.46 Embedded data

4 For a linked chart, you will open the original spreadsheet in a program such as Excel and can use all the facilities for formatting or editing the data.

5 Return to Word by closing the application.

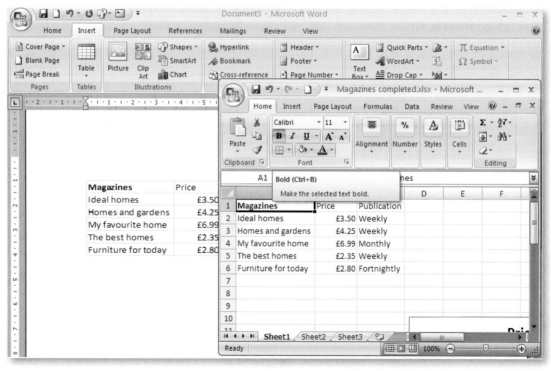

Fig. 1.47 Linked data

delete an imported object

1 Select the object with one click.

2 Press the **Delete** key.

resize an imported object

1 Click on the object once to select it.

2 Move the pointer over a corner (to maintain its proportions) until it shows a black double arrow and then hold down the mouse button.

3 Gently drag the boundary in or out. The pointer changes to a cross.

format an imported object

1 Click on the object and use normal alignment buttons to centre or right align it on the page.

2 Right click and select the **Format** option to open the dialog box or use tools on the **Home** and **Page Layout** tabs. Some of the changes you might want to make include setting an exact size, adding borders and setting a text wrap for text wrapping round it on the page.

3 Note that some options may not be available for imported objects.

4 If a text wrap is set, you will be able to drag the object to a different position on the page.

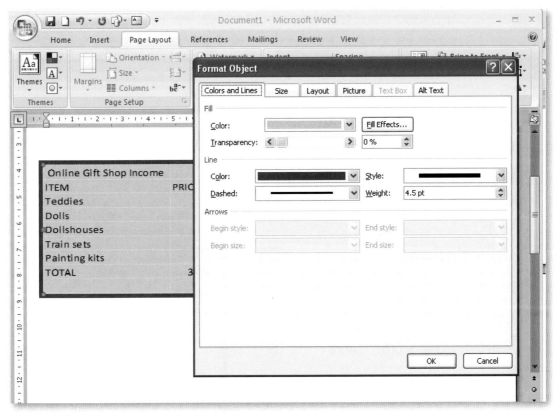

Fig. 1.48 Format an imported object

continue working below imported objects

1 Click on the imported object to select it.

2 Click again and the cursor will appear to its right.

3 Press **Enter** to move the cursor onto a new line.

4 Continue typing as normal.

or

5 Double click on the page below the object to display the cursor.

convert imported text into a table

1 Double click on the text you want to convert if it cannot be selected with the mouse.

2 Select the text and click on the **Insert** tab.

3 Click on the drop-down arrow below the **Table** button and select **Convert Text to Table**.

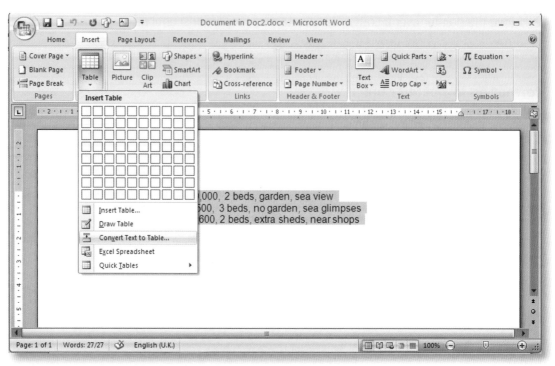

Fig. 1.49 Convert text to table1

4 In the dialog box that opens, select the number of columns you want the table to have and, if necessary, click on the checkbox to change the separator. An appropriate option such as paragraphs, commas or tabs will have been selected automatically.

Fig. 1.50 Convert text to table2

5 Click on **OK** and the table will appear.

There are two different ways you can line up blocks of text across the page:

- using tabs
- creating tables.

Check your understanding 11

1 Start a new document and save as *Problems*.

2 Type the following:

> **This year we saw a range of problems with the new computer system. After logging, they were all dealt with satisfactorily. The main problems were:**

3 Leave a clear line space and then import the data from *2009 problems*.

4 Centre align the data on the page.

5 Update *Problems* to save this change.

6 Close and then reopen *Problems*.

7 Now amend the data so that Harry becomes **George**.

8 Widen the display of the data.

9 Add a coloured fill to the heading row.

10 Save as *Problems resolved* and close the file.

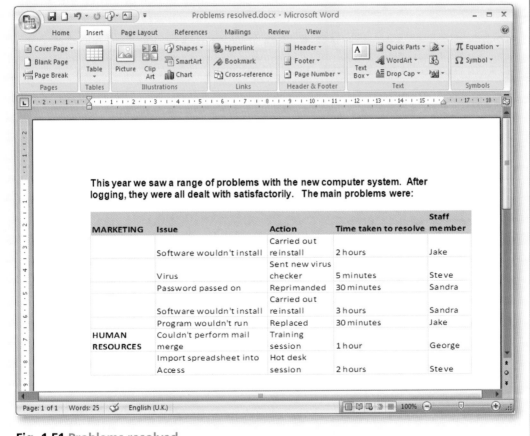

Fig. 1.51 Problems resolved

Importing charts and graphs

There are three different ways to add a chart to a Word document:

- Copy and paste a chart saved elsewhere, for example in an Excel file.
- Insert a data file that contains a chart on a separate sheet.
- Create a chart from scratch. The process is similar to using the wizard to create a chart within Excel.

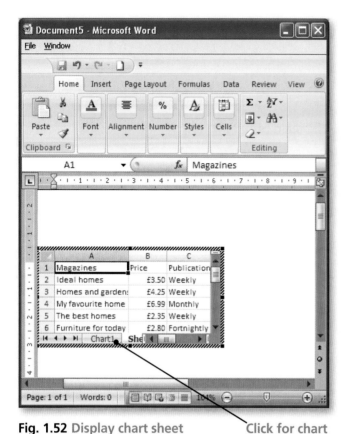

Fig. 1.52 Display chart sheet   Click for chart

insert a chart or graph already created

1 Click on the page where you want the chart to appear.

2 Click on the **Insert** tab and then click on the drop-down arrow next to **Object**.

3 Click on **Object** and then click on the **Create from File** tab.

4 Click on **Browse** to locate the data file such as an Excel file containing the graph.

5 Decide whether or not to create a link with the file so that changes are updated automatically. Then click on **OK**.

6 When the data appears on the page, double click for the sheet tabs and scroll bars. You may need to navigate to the chart if it is not on Sheet 1.

7 Use the chart tools to edit or format the graph. (See Unit 002 on spreadsheets for full details.)

8 Back in Word, click on the chart and drag out a boundary from a corner if you want to change its size.

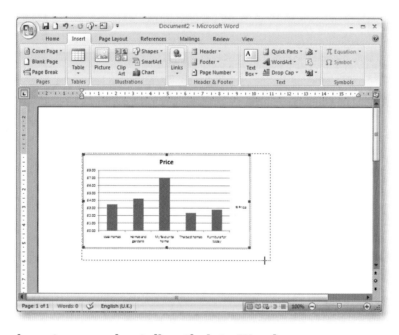

Fig. 1.53 Resize imported chart

insert a new chart directly into Word

1 Click on the **Insert** tab and click on **Chart**.

2 Select the type of chart, for example 2D column, from the range presented.

3 A ready-made chart will appear together with some data already entered in a spreadsheet. This window will be labelled *Chart in Microsoft Office Word*.

4 Start replacing the data with your own in the spreadsheet. The chart will be amended automatically on the page.

5 If you need further columns or rows, drag the lower, right-hand corner of the spreadsheet data area to make room for your entries.

6 When all the data has been added, close the chart to return to your page.

7 Click on Word's **Chart Tools – Layout** tab to add axes and chart titles.

8 Reopen the spreadsheet data at any time by clicking on the **Edit Data** button.

Chart reflecting changes to data   Reopen spreadsheet data   Chart tools

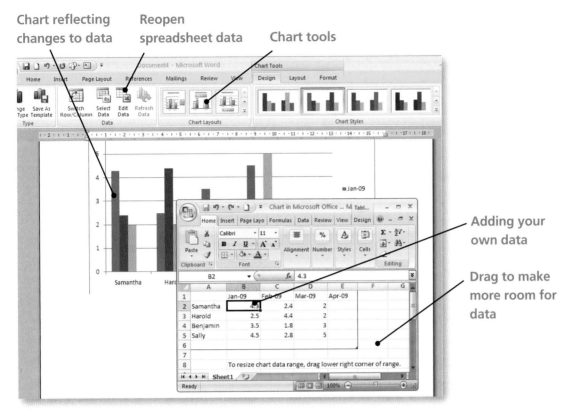

Adding your own data

Drag to make more room for data

Fig. 1.54 Create chart

Check your understanding 12

1 Start a new document.

2 Type the following:

<u>New Soups</u>

Here are the sales figures for our seven new soups. As you can see, Greek Salad was the favourite, closely followed by Mediterranean.

3 Now import the chart from the file *Soups*.

4 Make sure it is large enough for all the data to be seen clearly but does not extend into the page margins.

5 Centre the chart on the page.

6 Add the following text under the chart, making sure you leave at least two clear line spaces:

Many thanks to all the team who made this success possible.

7 Save as *New Soups with chart* and close the file.

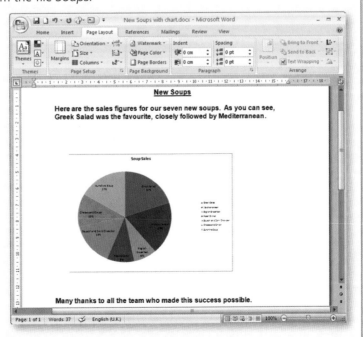

Fig. 1.55 New soups with chart

Importing images and graphics

For many types of document, you will want to include one or more images. These can be from a variety of sources:

- a digital camera
- a scanner
- files previously saved on your computer
- a CD-ROM
- Microsoft's Clip Art gallery
- the Internet.

Pictures are inserted in different ways, depending on their origin.

Clip Art

Although there is a reasonably wide range of pictures organised into categories available within Word 2007, many people will go online because of its much wider range and variety. Remember that many pictures are copyright and cannot be used for commercial purposes.

insert a Clip Art picture

1 Click on the **Insert** tab.

2 Click on **Clip Art** to open a search box to the right of the screen.

3 Type a brief subject, for example *buildings*, in the search box. (You could restrict your search to particular media files such as photos before searching by amending the ticks in any checkboxes in the **Results should be:** box.)

4 Click on the **Go** button and relevant pictures will appear.

5 Scroll down until you find a picture you like.

6 Click on it once and it will appear on your page.

Click for search pane

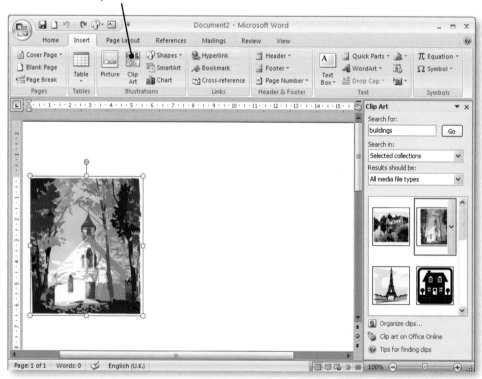

Fig. 1.56 Insert Clip Art

The picture will appear selected, showing a border with white squares and circles round the edge (sizing handles). To take off the selection, click in a white space on the page. To reselect it, click on the picture once.

Saved pictures

If you have saved a picture or have one available on a CD, you can find it and insert it into your document. It can then be edited in the same way as a Clip Art image. The common file types you will come across are:

- **Bitmap** or **raster images** (comprising hundreds of dots of colour called pixels).

> **TIF** – a common format for scanned images
>
> **JPEG** – compressed image files that offer a good range of colours
>
> **GIF** – simpler images
>
> **BITMAP** – large files created using Microsoft Paint

- **Vector images** – these are made up of a collection of curves and lines. They can be scaled down without loss of quality as this is not limited by the number of dots per inch. Common file types include:

> **EPS** (Encapsulated PostScript)
>
> **WMF** (Windows Metafile)
>
> **AI** (Adobe Illustrator)
>
> **CDR** (CorelDraw)

insert a picture from file

1 On the **Insert** tab, click on **Picture**.
2 This opens up the folders structure of your computer.
3 Navigate to the folder containing the picture. It may be in the **My Pictures** folder already created inside **My Documents**, or you may have to go up a level, for example to look in **My Computer** for a CD-ROM in the D: drive.
4 Once you can see the picture you want to insert, click on it in the main window to select it.
5 Click on the **Insert** button.

Select the picture   Click to open Insert
Picture dialog box

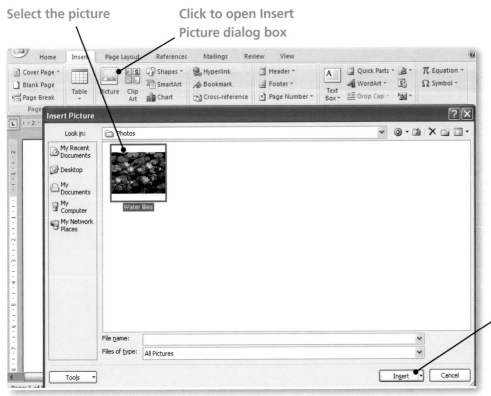

Click to add to document

Fig. 1.57 Insert picture
from file

6 When searching for a picture, you can set the computer to display thumbnails or a preview of any image files instead of just their names or details. Do this by selecting an alternative option from the **Views** button.

Views button

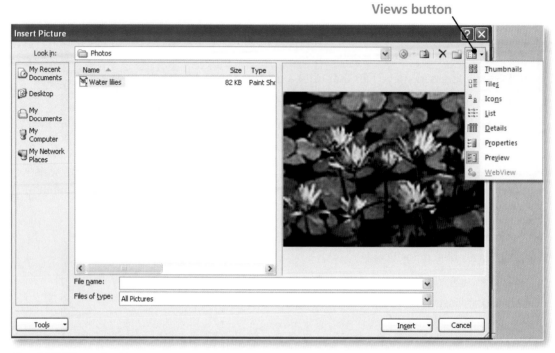

Fig. 1.58 Views button

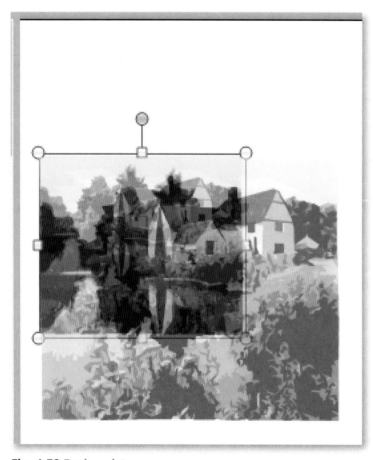

Fig. 1.59 Resize picture

Editing images

Once on the page, you can make a variety of changes to the images.

resize a picture

1 Click on the picture to select it.

2 Position the pointer over a corner sizing handle.

3 When it changes to a two-way arrow, hold down the mouse button and gently drag the pointer inwards or outwards.

4 The pointer becomes a small cross and the newly sized picture will appear as a faded image.

5 Let go when the picture is the correct size.

Note that you should use corner rather than central sizing handles to maintain the picture's proportions.

resize a picture exactly

1 Select the picture and click on the **Picture Tools** tab.

2 Amend the height and/or width measurements in the **Size** boxes.

3 You can also click on the arrow to open the **Size** dialog box. Here, amend measures and, to keep the picture in proportion, make sure there is a tick in the **Lock aspect ratio** box.

4 To increase or decrease the overall size by a percentage, change the setting in the **Scale** boxes.

move a picture

1 Position the pointer over the picture.

2 When the pointer shows a four-way arrow, hold down the mouse button and drag the picture across the screen.

3 The pointer will show a white arrow ending in a small box and you will be able to move the picture in a limited way, e.g. before or after text on the page. Its new position will show as a dotted, vertical bar.

4 You can also use the alignment buttons on the **Home** tab to centre or realign the picture on the page.

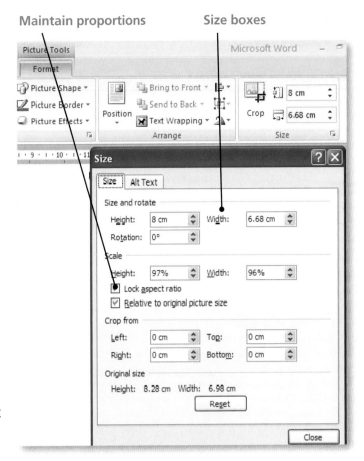

Fig. 1.60 Exact picture size

Using Text Wrap

It is more flexible to be able to drag a picture round the page, so that it can be positioned accurately. To do this, you need to change its properties. This is done by applying a text wrap. Once applied, the picture can be dragged around with the mouse.

Text wrapping is also used to control how text wraps round an object, for example when an image is positioned alongside text within a column.

apply a text wrap

1 Select the picture.

2 On the **Picture Tools** tab, click on **Text Wrapping**.

3 Select an option such as **Tight** or **Square**.

4 You will now find you can drag the picture easily.

5 You can also drag the picture up into a block of text and set how the text flows round it.

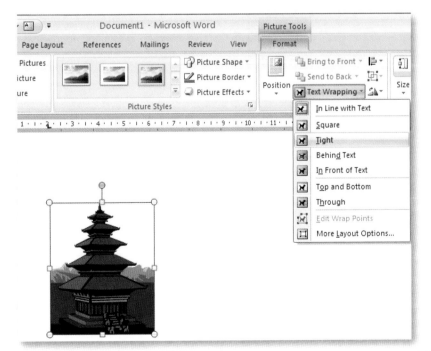

Fig. 1.61 Text wrap picture

delete a picture

1   Select the picture you want to delete.

2   Press the **Delete** key.

Check your understanding 13

1   Open the file *Antiques* on the CD-ROM accompanying this book.

2   Leave a clear line space and then insert a Clip Art picture of any household item after the text **talking about eBay**.

3   Reduce the size and make sure you keep it in proportion.

4   Centre it on the page.

5   At the end of the document, leave a clear line space and insert the picture *vases.jpg* from the file.

6   Reduce the size so that the whole document fits on one page.

7   Right align the picture on the page.

8   Save the file as *Antique pictures*.

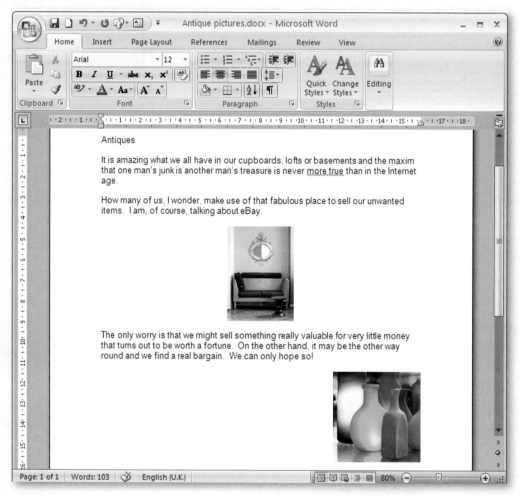

Fig. 1.62 Antique pictures

Sections

When compiling long documents that incorporate imported material such as charts or tables of data, it may be necessary to retain the integrity of the additions in terms of their orientation, margins, headers and footers and so on. To do this, you need to divide the document into sections. Each section can then have its own layout.

create sections

1 Click in place for the next section.

2 On the **Page Layout** tab, click on **Breaks**.

3 Select **Section Breaks**.

4 You can select the option to start the next section on a new page or for it to continue from this point.

5 You can now make layout changes to this section of the document only.

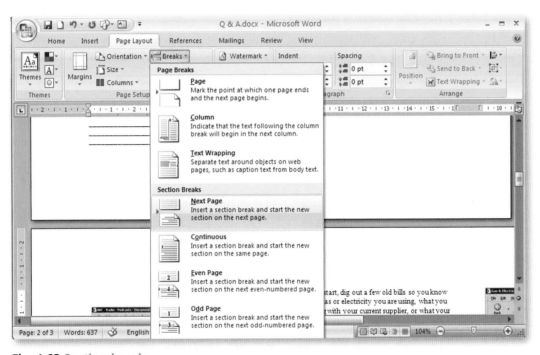

Fig. 1.63 Section break

Setting text in columns

There are two different ways you can line up blocks of text across the page:

- using tabs
- creating tables.

Using tabs

Each time you press the **Tab** key (labelled with two arrows, next to the letter Q) it moves the cursor across the page. The normal distance it moves is in jumps of 1.27cm (0.5 in). If you fix the position it moves to (known as the **Tab stop position**), you can use this function to help set out information in a document in neat columns. You do this by placing black symbols along the ruler exactly where you want the cursor to move to for each column.

As well as the tab stop position, you can also set the *way* in which columns line up (**Tab alignment**):

- by their initial characters (**Left Tab**)
- by their last characters (**Right Tab**)
- centred on the tab position (**Centre Tab**)
- with decimal places in a column of figures lining up exactly (**Decimal Tab**).

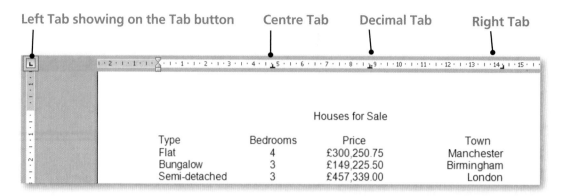

Left Tab showing on the Tab button   Centre Tab   Decimal Tab   Right Tab

Fig. 1.64 Columns with tabs

fix tab stops using the ruler

1 Make sure the horizontal ruler is showing.

2 Start by typing any column headings, spacing these by eye. It will be easier to get their position correct after typing all the text if they are *not* set using the **Tab** key.

3 Press **Enter** to position the cursor on the left margin.

4 Set the position for the first column. If you do not want a Left Tab, first click on the **Tab** button above the left scroll bar until it shows the correct type of tab symbol, for example **Right** or **Centre**, then click on the ruler where you want the cursor to move to. The chosen symbol will appear.

5 Repeat the process until all the tab stops are on the ruler.

6 If necessary, click on the **Tab** button again for a different alignment.

7 If you make a mistake, rest the pointer on the tab stop and then gently drag it up or down away from the ruler. When you let go of the mouse the symbol should disappear. You can also drag a tab stop to a different position along the ruler.

8 To type your document, either type the first entry so that the first column lines up on the left margin or press the **Tab** key and start typing where the cursor moves to. Press the **Tab** key again to move the cursor to the next column and repeat across the page.

9 When you have completed the first row, press **Enter** to move to the next line and start typing the next row of entries.

10 If you find you need a wider column, select **all** the text typed using tabs. Now drag the position for the incorrect tab along the ruler. A dotted line will show its progress and the column position will be readjusted.

use the menu

1 Type the headings and then press **Enter** to move to the first line for the column entries.

2 Type the first entry and then press the **Tab** key once to move to the next column. Repeat across the page. Your text will appear very squashed but will be sorted out later.

3 Repeat for the remaining entries, pressing **Enter** each time to move to the next row. Highlight all the column entries set with tabs.

Type	Bedrooms	Price	Town
Flat	4	£3550000	London
Maisonette	3	£127,580	Manchester
House	5	£467890	Birmingham
Semi-detached	4	£334560	Exeter

Table 22.2 Single tab between entries

4 On the **Home** tab, click on the arrow to open the **Paragraph** dialog box and click on the **Tabs** button.

5 In the **Tabs** dialog box, click in the **Tab stop position** box and enter the measure (figures only, no cm) for the first column, for example Bedroom entries would be positioned in a column lined up at 4cm along the ruler by entering **4**. Click for **Centre**, **Right** or **Left** style of tab and then click on the **Set** button.

6 Click in the box again (you will have to type over the entry that appears) and enter the position and tab alignment style for the next column. Repeat until all tab stop positions have been set.

7 Return to your document and check that the columns look right. If not, return to the dialog box and change any measurements or tab styles.

8 To remove unwanted tab stops, select them in the box and click on the **Clear** button.

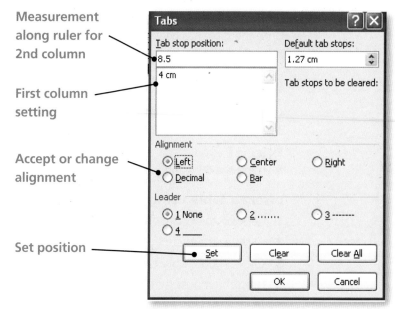

Measurement along ruler for 2nd column

First column setting

Accept or change alignment

Set position

Fig. 1.65 Set tabs with menu

Check your understanding 14

1 Start a new document.

2 Enter the title *Animals for Food*.

3 Save the file with the same name.

4 Centre the title and apply uppercase.

5 Now create the table set out below. Enter the column headings first and then use tabs.

6 Set the columns as follows:

- Type – on the left margin
- Meat – Left tab at 3cm
- Price per kilo – Decimal tab at 7cm
- Varieties – Right tab at 13cm

Type	Meat	Price per kilo	Varieties
Pig	Pork	£8.45	Bacon, Ham
Cow	Beef	£12.75	Veal
Lamb	Mutton	£9	Lamb

7 Format the column headings in bold and italic and make sure they are positioned correctly over the column entries.

8 Finally, move the 'Price per kilo' column to 8.5cm and adjust the headings so that they are still over the column entries.

9 Save and close the file.

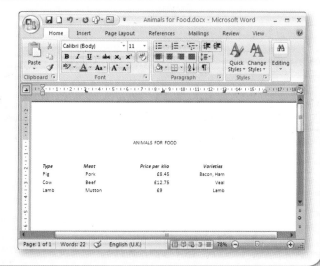

Fig. 1.66 Animals for Food

Working with tables

Tables are a series of rows and columns of cells. Each cell contains discrete data that can be aligned, formatted and edited separately. Single rows and columns or the whole table can have visible gridlines; you can add emphasis in the form of enhanced lines and shading or, by removing the gridlines, the data can be displayed in simple columns.

Once the table has been created and you want to start entering the data, move from cell to cell using the **Tab** key or mouse.

Creating a table

There is an option available to draw your own table if it has a complex shape, but normally you use one of two methods to set up the correct number of columns and rows:

- using the mouse
- using the table menu.

insert a table using the mouse

1 Click on the page where you want the table to appear.

2 Click on the **Insert** tab.

3 Click on the **Table** button.

4 Drag the mouse across the rows and columns of cells to create your preferred size of table. You will see it taking shape on the page.

5 Let go of the mouse and you will return to your page.

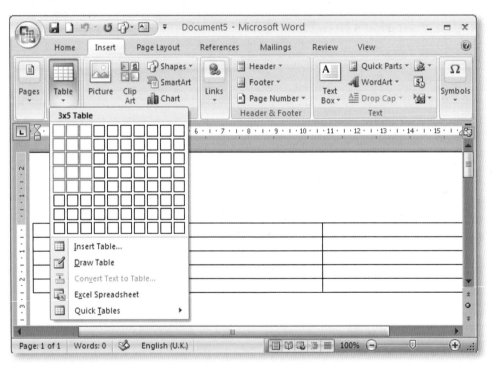

Fig. 1.67 Insert Table

Fig. 1.68 Insert Table menu

insert a table using the table menu

1 Click on the **Insert Table** option.

2 Enter the required number of columns and rows.

3 Click on **OK** and the table will appear.

amend columns and rows

1 Click in the last cell and press the **Tab** key to add a new row of cells.

2 Right click any cell, select **Insert** and select the appropriate option such as a new column to the right or a row above.

3 Right click on a cell, select **Delete cells** and then select an option such as an entire row or column.

4 You can also click on the **Layout** tab under **Table Tools** and click on an **Insert** or **Delete** option.

5 If you click on the small square in the top left-hand corner of the table and select the whole table, pressing the **Delete** key will delete cell contents rather than the actual table.

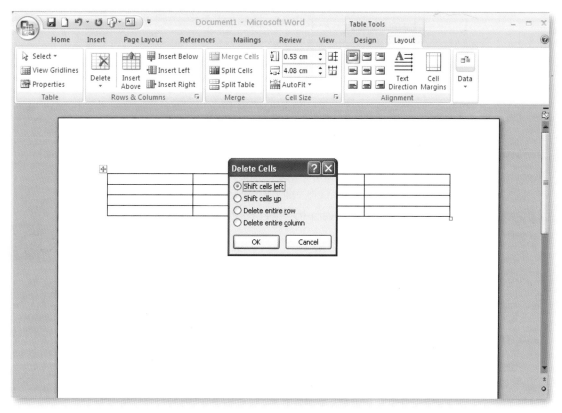

Fig. 1.69 Delete table

Column widths and height

The default size for each table cell depends on the number of columns in the table and the entries you have made. If the default measurements do not allow you to display your data appropriately, you can amend individual cell sizes or that of an entire column or row.

amend measurements manually

1 To change column width, move the mouse up to the ruler and hover over the blue marker showing the right-hand column boundary.

2 When it displays a black two-way arrow, click and hold down the mouse button.

3 Gently drag the border to the right to increase the width of the column.

4 You can also drag a column border with the mouse from within the table.

5 If you double click on the cell boundary within a table, it will set the width to the longest entry.

6 For row height, follow the same process but drag the boundary vertically up or down.

7 If you double click on the ruler, you will open the **Table Properties** box and can make other changes such as setting measurements for each row or column exactly.

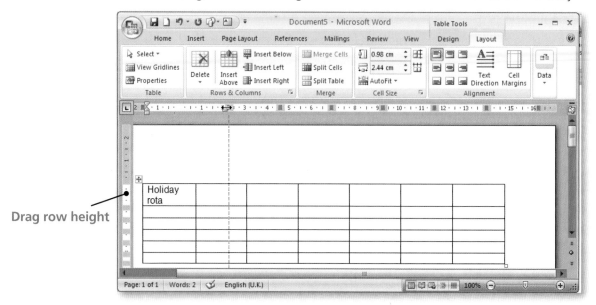

Drag row height

Fig. 1.70 Drag table cell width

amend cell measurements using the Table Tools

1 Select the target column by moving the pointer above it until it shows a down-facing black arrow. Click and the column cells will turn blue.

2 Click and drag the pointer if you want to select more than one column to adjust at the same time.

3 To select rows, click and drag the cell contents or click and drag the pointer in the left margin when it shows a right-facing white arrow.

4 Click on **Layout** under **Table Tools** on the ribbon.

5 Click on the arrows in the **cell width or height** box to increase or decrease the measure, or enter your own figures and press **Enter**.

Row height   Column width measure

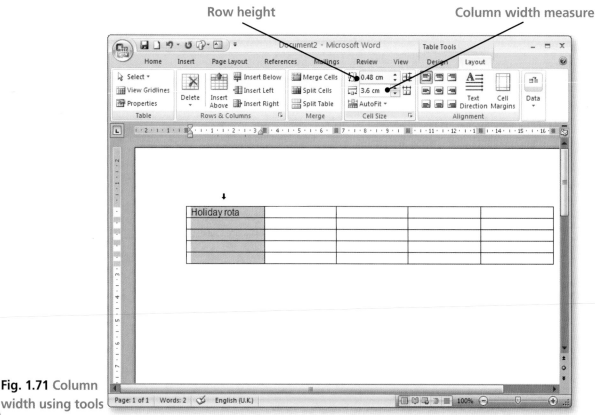

Fig. 1.71 Column width using tools

Cell alignment

Entries in cells can be aligned vertically or horizontally if you want to change how the data is displayed.

Horizontal

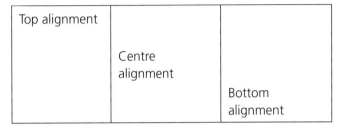

Left aligned entry	Centre aligned entry	Right aligned entry

Vertical

Top alignment		
	Centre alignment	
		Bottom alignment

You can also combine the alignments, for example **Bottom Left** or **Top Right**, so there are nine different alignments that can be set.

set cell alignment

1 Select the target cell(s).
2 Right click and select **Alignment**.
3 Click on the appropriate button.
4 These can also be found on the **Layout** tab under **Table Tools**.
5 You can also set horizontal alignment using the **text alignment** buttons on the **Home** tab.

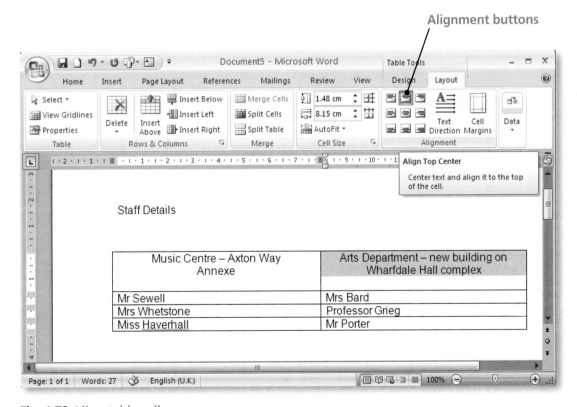

Fig. 1.72 Align table cell

Check your understanding 15

1 Start a new document and save as *Beads*.

2 Enter the title **Beads**. Set the font size to 20 and centre the heading on the page.

3 Create a table that has four columns and five rows.

4 Enter the following four headings:

   Type of bead   Main colour   Size   Cost per pack of 10 beads

5 Now add the following data:

Type of bead	Main colour	Size	Cost per pack of 10 beads
Lampwork	Multi	8mm	£5
Rocaille	Gold	1mm	£3.50
Bugle	Silver	4mm	£4
Bicone	Blue	10mm	£8.25

6 Set the heading row height to 1.5cm.

7 Increase the font size for the heading text to 16 and the main entries to 14.

8 Centre align the column headings at the top of the row.

9 Change the width of the final column if necessary so that the heading is on three lines.

10 Realign the first three headings so they are centred vertically as well as horizontally.

11 Centre the table on the page. Either use the **alignment** buttons on the **Home** tab or move the table by dragging the move handle in the top left-hand corner showing four arrows.

12 Save the file to update these changes.

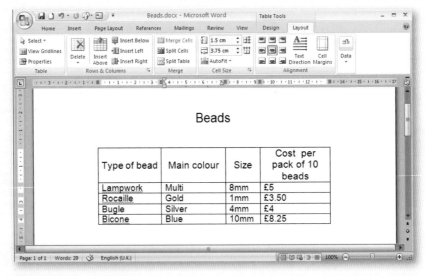

Fig. 1.73 Beads table

Merging or splitting table cells

To be able to centre a heading above several columns in a table, you will need to merge the cells. You can also split a cell if you want two or more entries in the original space.

merge cells

1 Select the cells you want to merge.

2 On the **Table Tools – Layout** tab, click on the **Merge Cells** button.

3 The selected cells will now become a single cell.

4 Format the cell contents in the normal way.

Two cells already merged   Merge button   Split cells   Selected cells

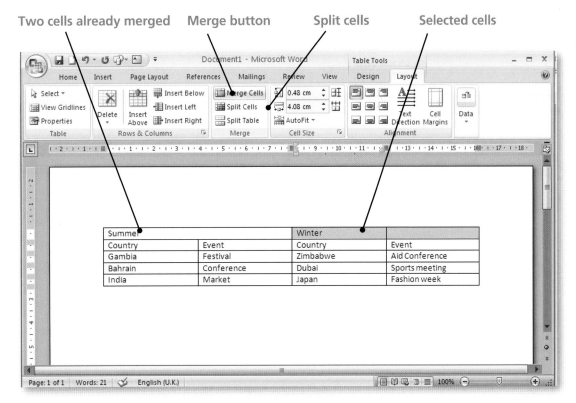

Fig. 1.74 Merge cells

split cells

1 Select the cell(s).
2 Click on the **Split Cells** button on the **Table Tools – Layout** tab.
3 Select the number of columns and/or rows to create.
4 Click on **OK**.

Using tabs in a table

In some cases, you may want to indent table entries within a cell. You can move across a cell in jumps of 1.27cm (0.5 in) when using the **Tab** key, but to indent entries at an exact point on the ruler, or to set the way they line up, you will need to set tab stops at the position you want the cursor to move to.

As you move from cell to cell if you press the **Tab** key on its own, you need to hold down **Ctrl** to move to a tab stop set *within* a cell unless it is a Decimal tab.

set tabs within a table

1 Click on the cell in which you want the tab stop. For the same tab stop for an entire column, select all the column cells first.
2 Click on the **Tab** button to the left of the ruler until the correct style of tab is showing – cycling through left, centre, right and decimal.

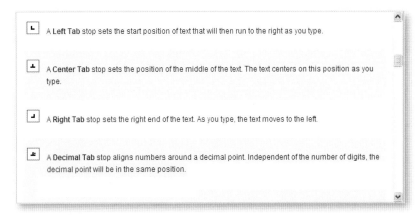

Fig. 1.75 Tab styles

3 Click on the ruler for the tab stop position.

4 Select the next cell/column and repeat.

5 Do this for all the tab stops you want in the table.

Tab button

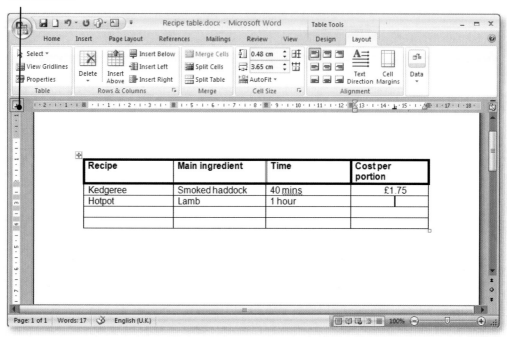

Fig. 1.76 Tab set

use tabs in a table

1 Click in the first cell and enter any data as normal.

2 Move across the table and to the cell containing the tab stops by pressing the **Tab** key.

3 For decimal tabs, type in the figure. The cursor automatically moves to the tab stop position.

4 For other types of tab, hold **Ctrl** and press **Tab** to move the cursor to the tab stop position within the cell.

5 Continue to enter data into the table in the normal way.

Left tab stop in cell

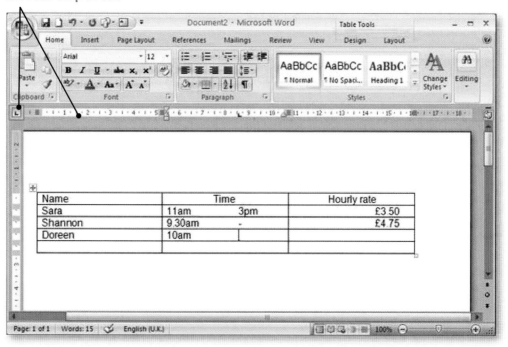

Fig. 1.77 Using tabs in a table

Check your understanding 16

1 Start a new document and create a table with four rows and three columns.

2 You will be entering the data as shown below.

3 You want to indent the names. Set a **Left** tab in the first column 1cm from the cell boundary.

4 Set a **Left** tab in the second column 2cm from the cell boundary to indent the course data.

5 Enter column headings without using tabs, and centre the headings in the first row.

6 In the first cell, move to the tab stop and enter the text **Sally**.

7 Move to the next cell. Type the two lines of text. (To add entries on a new line within the same cell, press **Enter**.)

8 Move to line 3 and then move the cursor to the tab stop position to enter **Advanced Word**.

9 Move to the next line and type **Web pages** at the tab stop position.

10 Use the **Tab** key to move to the next cell to enter the grade.

11 Enter the rest of the data, lining up entries as shown.

12 Save as *Training table.*

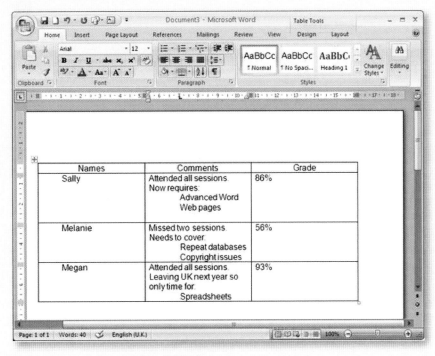

Fig. 1.78 Training table with tabs

Borders and shading

To add more emphasis to a table, you can change the width and style of cell border and colour and shade both the lines and the cell backgrounds. Alternatively, you can remove borders to display just the data in columns on the page.

change borders

1 Select the target cells.

2 Click on the **Table Tools – Design** tab.

3 Click on the top **Line Style** button drop-down arrow to change from a single line to double lines or dashes, and the **Line Weight** arrow to select a line thickness.

4 Click on the **Borders** button and select an option such as **Outside Border** to apply the border.

5 If you change line style or weight, remember to click on the **Borders** button again to apply the new settings.

6 Remove borders by selecting **No Border** from the Borders or Line Style menus.

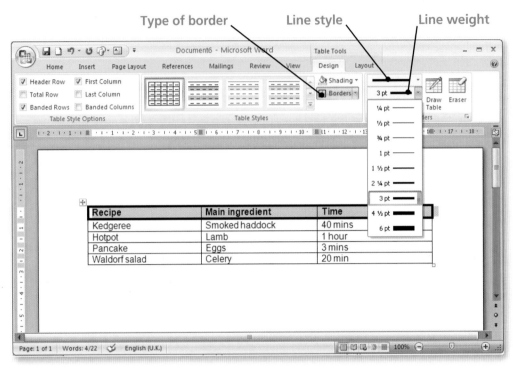

Fig. 1.79 Border buttons

or

7 Right click in the table and select **Borders and Shading** to open the dialog box. This option is also available from the small arrow in the bottom corner of the **Draw Borders** box on the ribbon.

8 You can now select border style, weight and type from the various boxes in the same way that you did for border text.

9 To remove a partial border such as a horizontal top border, click on the correct button in the **Preview** section.

10 You can also change the colour of the border line.

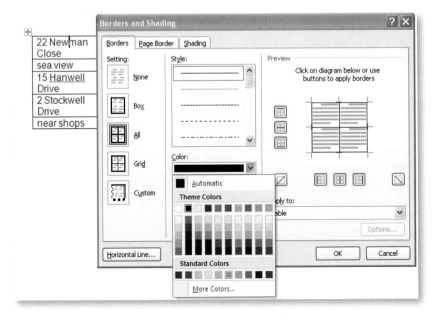

Fig. 1.80 Borders dialog box

shade cells

1 Select the target cells.

2 On the **Table Tools – Design** tab, click on the drop-down arrow in the **Shading** box to select from a range of different colours.

3 You can preview the effect by resting your mouse on the selected colour.

4 If you choose from the themes rather than standard colours, the appearance of the fonts you are using may also change.

5 Click on **More Colors** for a wider palette.

6 Remove any shading by selecting **No Color**.

7 You can also apply a pre-set table style offering different borders and shading by clicking on one of the examples on the ribbon.

Table styles

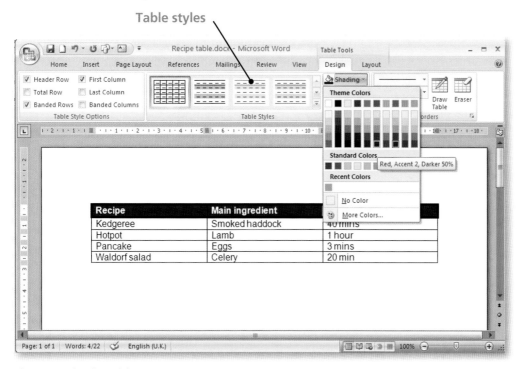

Fig. 1.81 Shade table

Check your understanding 17

1 Reopen *Beads*.

2 Add a thick, dark border to the outside of the entire table.

3 Shade the heading row mid-blue.

4 Shade the bead names pink.

5 Save as *Shaded beads*.

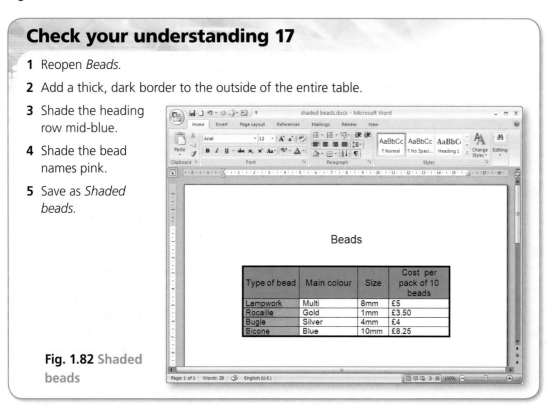

Fig. 1.82 Shaded beads

Page layout

At Level 1, you learned how to enter, format and edit simple documents. You should be able to change font, font size and font colour and apply a text alignment such as centred or justified. These options are available on the **Home** tab, from the relevant dialog box and also from a floating toolbar offering a limited range of tools that appears in Word 2007 whenever text is selected.

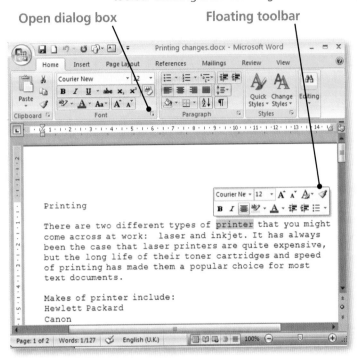

Fig. 1.83 Floating formatting toolbar

You should also be able to delete or insert entries and move or copy text or objects. Here is a reminder of the steps to take when moving or copying.

move or copy text

1 Select the word(s) or block of text.
2 Click on the **Cut** icon to move or the **Copy** item to copy. These are available by right clicking or on the **Home** tab in the Clipboard area.
3 Click on the page or open the document where the text is to appear.
4 Click on the **Paste** icon.
5 If necessary, adjust any spacing.

At Level 2, the emphasis is more on using some of the advanced features of the software to create and edit long documents and make sure they are accurately and professionally laid out.

The main aspects of document layout that you need to know how to change include line or paragraph spacing, margins, page orientation and page and other breaks in the text. (To make changes to a whole document, select it quickly by holding down **Ctrl** and pressing the key **A**.)

House style

Most organisations have a house style by which their documentation can be recognised. Usually this involves using a specific type of font or range of font colours, specified page margins and a certain type of border for images or logos. They may also prefer serif fonts such as Times New Roman or sans serif fonts such as Calibri. For example:

● all headings might have to be Arial, font size 16, red
● all subheadings might have to be Arial, font size 14, black

When asked to edit a document and apply the house style, it can take time to change the formatting for each heading and subheading. A quick way to do this is to use the Format Painter utility to literally *paint* a format onto selected text.

use Format Painter

1 Select any part of the text that already has the correct formatting applied, or format some text first and then select this.
2 Click on the **Format Painter** button on the **Home** tab. (To repeat the formatting throughout a long document, double click on the button to keep it turned on.)
3 Use the scroll bars if necessary to display the first entry where the new format needs to be applied. (If you click on the page you will cancel the process or format the wrong text.)
4 Gently drag the pointer along the block of text. The pointer will display a small brush.
5 When you let go of the mouse, the text will have taken on the copied formatting.
6 If necessary, continue throughout the document and then click on the **Format Painter** button to turn it off.

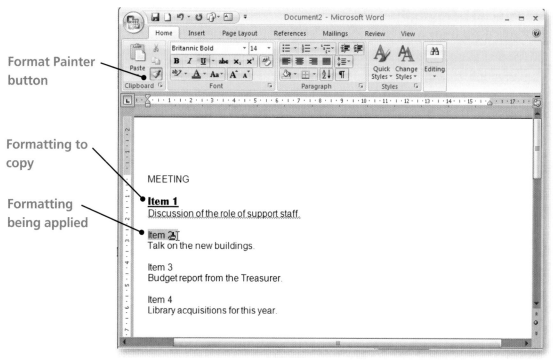

Format Painter button

Formatting to copy

Formatting being applied

Fig. 1.84 Format Painter

Paragraph and line spacing

As well as adjusting the spacing between individual lines in a paragraph, you can set the amount of white space left before or after whole paragraphs or selected headings or subheadings.

For the assessments, you will be asked to set particular spacing options. First make sure that no special spacing options have been applied by default or have accidentally been set.

check paragraph settings

1 Select the document.
2 Right click and select **Paragraph** or click on the small arrow in the **Paragraph** section on the **Home** tab to open the **Paragraph** dialog box.
3 Check that the entry in the **Line spacing** box is *Single*.
4 Check that the left and right indentations are set at 0.
5 Check that there is no **Special paragraph** set.
6 Check that the **Before:** and **After:** boxes in the **Spacing** section show 0.
7 Make any changes as necessary.
8 Click on **OK** to close the box.

It is very useful, if you find normal text entry difficult, to check the settings in a range of dialog boxes such as **Font**, **Paragraph** or **Page Setup**. Often you will be able to remove an unwanted setting and then carry on with your work.

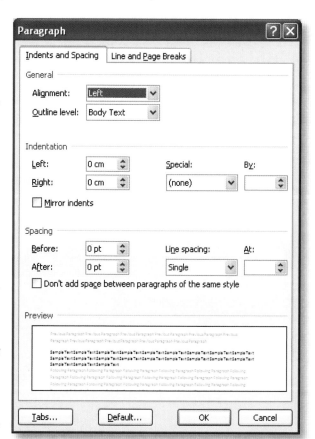

Fig. 1.85 Check paragraph spacing

set special paragraphs

1 Open the **Paragraph** dialog box.

2 For text to be indented in the first line but positioned on the left margin for the rest of the paragraph, click in the **Special** box and select **First line indent**.

3 For the first line to be aligned on the left margin but the rest of the paragraph indented, select **Hanging indent**.

4 In both cases, change the position for the indent by entering an exact measure in the **By:** box, or drag the indent marker along the ruler.

Hanging indent measure changed

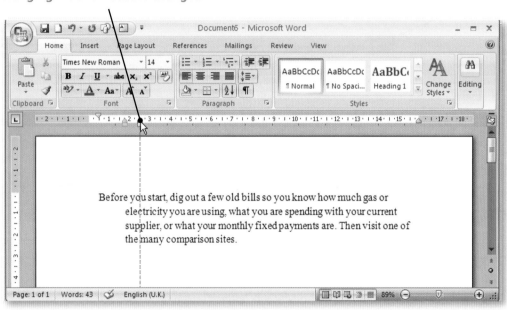

Fig. 1.86 Change hanging indent

Line spacing

The default line spacing is single – each line follows on from the one above with no spaces. This means that every time you press **Enter,** the cursor should move to the line below, whether it is at the end of a sentence within a paragraph or after a heading or subheading. To change this setting so there is white space between lines, you can set double, 1.5 or other line spacing.

Open dialog box

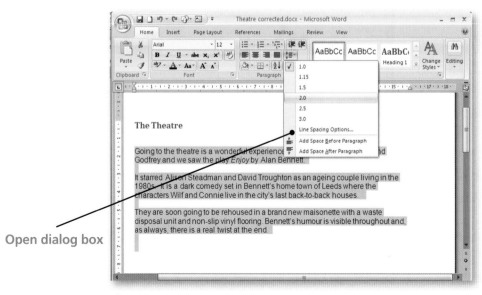

Fig. 1.87 Line space button

change line spacing

1 Select the paragraphs or whole document.

2 Use shortcuts to set:

- double line spacing (**Ctrl** plus **2**)
- 1.5 line spacing (**Ctrl** plus **5**)
- single line spacing (**Ctrl** plus **1**).

  or

3 Select options from the **Line spacing** button on the **Home** tab.

or

4 Right click or click on **Line Spacing** options to open the **Paragraph** dialog box.

5 Here you can choose from a range of line spacings or even set an exact measure. To do this, select an option such as **Exactly** from the line spacing drop-down list and enter the measure in the **At** box.

add line spacing before or after paragraphs

1 Select the heading or paragraph(s) to change.

2 Select a simple **Add Space** option from the **Line Spacing** button.

3 For a specific measure, open the **Paragraph** dialog box from the **Home** tab or by right clicking and selecting **Paragraph**.

4 In the **Spacing** section, click on the arrows or enter exact measures in the **Before:** or **After:** box.

5 The default measurement unit is points, but you can change to inches or centimetres if you add ' or cm in the box. Next time you look, these will have been converted back to the appropriate number of points.

6 Click on **OK**.

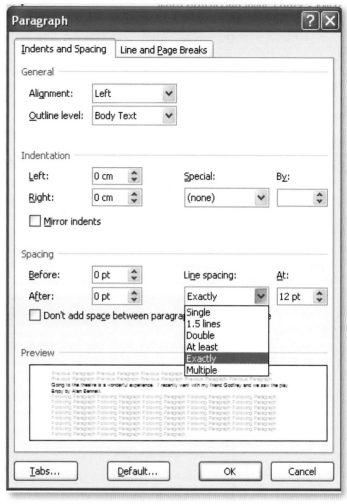

Fig. 1.88 Line space dialog box

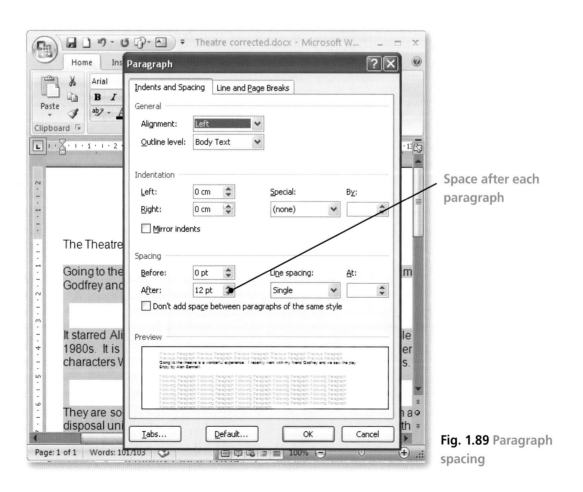

Space after each paragraph

Fig. 1.89 Paragraph spacing

Widows and orphans

The last line from a previous paragraph that starts on a new page is known as a **widow**, and the first line of a new paragraph that appears on its own at the bottom of the page is known as an **orphan**. These should be prevented from occurring in business documents. By setting the correct paragraph formatting option, single lines will be joined automatically to their appropriate paragraphs.

Note that it is normal to follow a heading or subheading with at least two lines of related text on the same page.

prevent widows and orphans

1   Select the paragraph or whole document.

2   Open the **Paragraph** dialog box.

3   Click on the **Line and Page Breaks** tab.

4   Check that there is a tick in the **Widow/Orphan control** box.

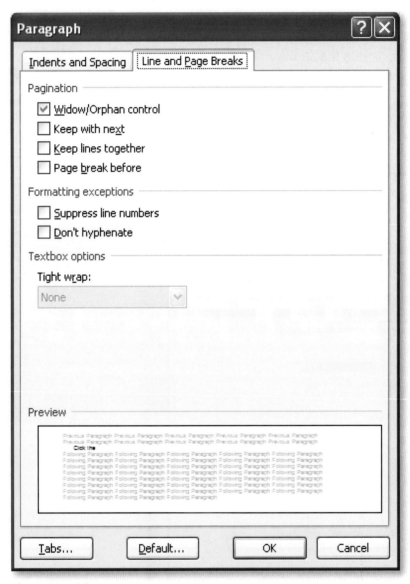

Fig. 1.90 Widows

Check your understanding 18

1 Open the text file *York* on the CD-ROM accompanying this book.

2 Format the entire document as follows:

- Font – Times New Roman
- Font size – 11
- Alignment – fully justified
- Emphasis – none
- Line spacing – single

3 Save as a Word 2007 document with the name *York formatted*.

4 Set the spacing after each paragraph to 6 pt.

5 Centre the heading and make it bold, font size 14.

6 Format the subheading *Location* as follows:

- Underlined
- Italic
- Font size 12

7 Format the four other subheadings in the same way.

8 Double space the paragraph headed *Viking Centre* only.

9 Save these changes and close the file.

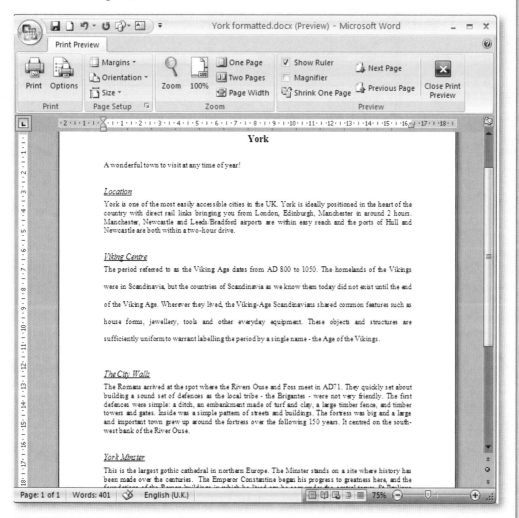

Fig. 1.91 York formatted

Margins

The normal Word 2007 document page margins are set at 2.54cm top, bottom, left and right. To change any of these measurements, you can either select a pre-set option or enter exact measurements into the margin boxes in the **Page Setup** dialog box.

(Note that in an assignment, there is a tolerance of 6mm.)

change margins

1 On the **Page Layout** tab click on the drop-down arrow below the **Margins** button.

2 Select one of the styles offered.

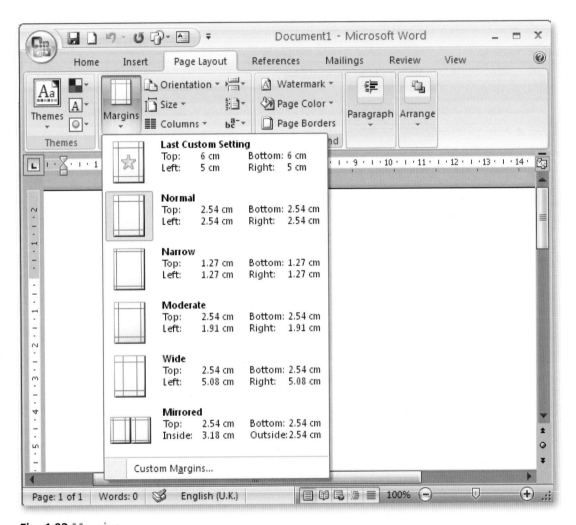

Fig. 1.92 Margins

3 To open the **Page Setup** dialog box, click on the **Custom Margins** link or the small arrow in the **Page Setup** group on the ribbon.

4 On the **Margins** tab you can use the up or down arrows or enter an exact measure in any of the margin boxes.

5 Note that 1/100th cm measures need to be entered manually into the margin boxes as you cannot set these using the up or down arrows.

Check your understanding 19

1 Open *York formatted*.

2 Change the left and right margins to 3.45cm.

3 Take a screen print to show these settings.

4 Update the file to save these changes.

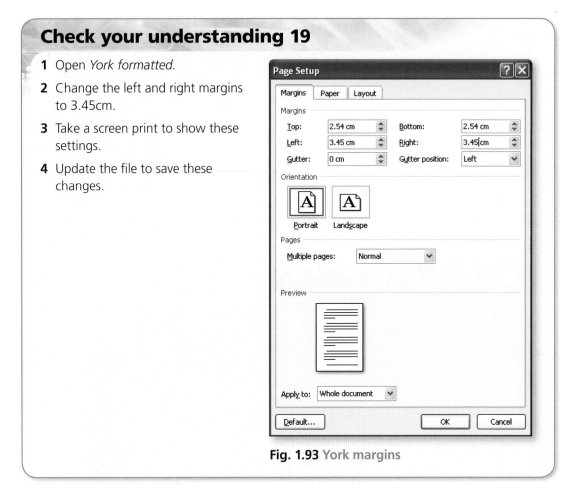

Fig. 1.93 York margins

Orientation

The default setting for Word 2007 is **Portrait** orientation where the shorter edges of the page are at the top and bottom. Some documents need to be **Landscape** to display the data more effectively with longer sides top and bottom.

change page orientation

1 Click on the **Orientation** button on the **Page Layout** tab and select the alternative.

or

2 Open the **Page Setup** dialog box and click on the correct button on the **Margin** tab.

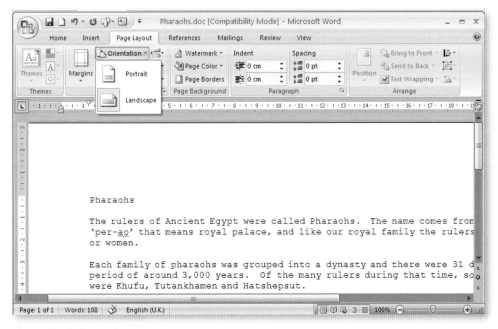

Fig. 1.94 Orientation

Page and paragraph breaks

If you want to break up a long paragraph into shorter ones, you can add a 'soft' paragraph break by clicking in front of the word that will start the new paragraph and pressing **Enter** twice.

By pressing **Enter** a number of times you could move any text that is to the right of the cursor onto a new page. However, if anything is added or removed earlier in the document, the position of the text will change. If you want a section of a document to always start on a new page, you need to set a 'hard' page break. This will not be compromised by changes elsewhere.

In Normal view, you cannot see the page break after it has been set. To see it, click on the **Show/Hide** button on the **Home** tab. This displays all keystrokes and special characters such as spaces between words, tabs and paragraphs. A page break shows as a labelled, dotted line.

Show/Hide button

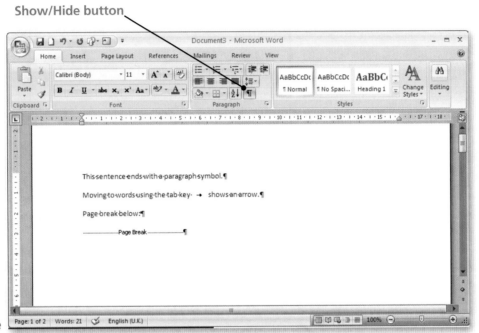

Fig. 1.95
Show/Hide

Insert blank page   Page break

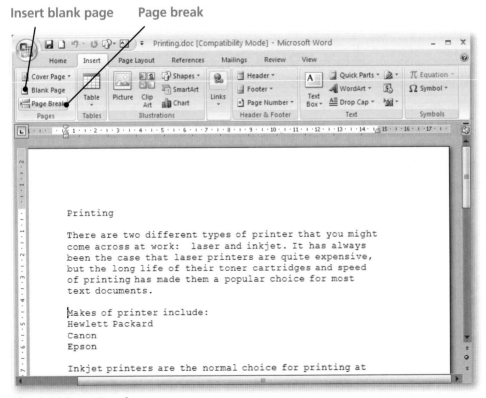

Fig. 1.96 Page Break

set a page break

1 Click in front of the text you want to start on a new page or click after the last paragraph.

2 Hold **Ctrl** and press **Enter**.

or

3 Click on the **Insert** tab.

4 Click on **Page Break**.

5 Make sure the text starts on the first line of the page and at the left margin unless other alignments have been set.

6 To separate the current text from the new page, you could insert a **Blank Page** at this point.

remove a page break

1 Click in front of the first word on the new page.

2 Press the **Backspace** key one or more times.

or

3 Click on the **Show/Hide** button, click on the dotted **Page Break** line and press the **Delete** key.

4 You may need to readjust spacing when the text moves back a page.

Check your understanding 20

1 Open the file *Printing* on the CD-ROM accompanying this book.

2 Create a new paragraph beginning **Many manufacturers**.

3 Insert a page break after the list of printers so that the text **Inkjet printers are the normal choice** starts on a new page.

4 Save as *Printing changes* and close the file.

Editing the text

If you are asked to revise or finalise a document, there are many facilities in Word 2007 that make the task quicker and easier. These include finding and replacing entries, inserting special characters and symbols and adding header or footer entries that will update automatically.

Using search and replace

As well as carrying out a simple search for a specific entry, or replacing an entry with another throughout a long document, the search and replace tools in Word 2007 allow you to carry out quite sophisticated searches or make changes to a document.

You can:

- find part or all of a word or phrase
- match case – to find entries that are case sensitive
- look for formatting that has been applied
- locate words that sound similar
- use wildcards – to find entries where the symbol * represents unknown characters.

search a document

1 On the **Home** tab, click on **Find**. The button shows binoculars.

2 Enter the word or phrase you want to search for in the **Find what:** box.

3 Click on **More>>** (toggles with **<<Less**) to open the **Search Options** window if it is not visible.

4 Click in the checkboxes as necessary to restrict the search.

5 To look for words with a particular emphasis or paragraph formatting, click on the **Format** button and select from the menus available.

6 To look for keystrokes, page breaks and so on, click on the **Special** button.

7 To search from the point you have reached in the document, select **Down** or **Up** in the **Search** box.

8 Click on **Find Next** to locate the first matching entry and repeat to work through the document.

Direction

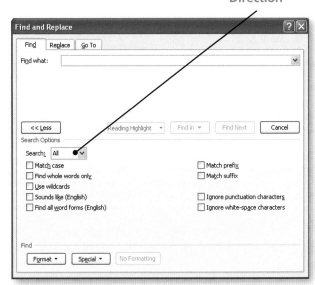

Fig. 1.97 Find

replace entries

1 Click on **Replace** on the **Home** tab (the button shows *ab → ac*) or click on the **Replace** tab in the **Find and Replace** window.

2 Enter the word or phrase in the **Find what:** box exactly as it appears in your document.

3 Enter the replacement entry in the **Replace with:** box.

4 To replace all entries, click on **Replace All**.

5 To locate and check an entry before replacing, click on **Find Next**.

6 To move on without replacing, click on **Find Next** again.

7 To replace a highlighted entry, click on **Replace**.

Check your understanding 21

1 Open the file *Italian* on the CD-ROM accompanying this book.

2 Insert a page break in front of the subheading *Rice.*

3 You need to replace entries of the word *rice* with *pasta.* Use the **Find and Replace** tool, but make sure you do not replace the subheading *Rice.*

4 Now carry out a search for an Italian sauce whose name you cannot remember. All you know is that it starts with *car* and ends with *a.*

5 Remove the page break you created earlier.

6 In the first paragraph, create a new paragraph beginning *This region...*

7 Save and close the file.

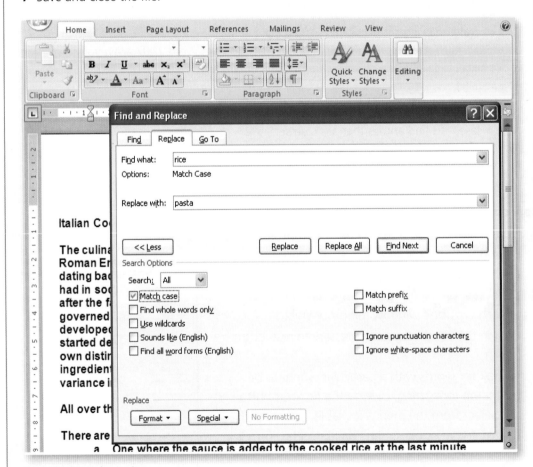

Fig. 1.98 Replace rice

Special entries

Some documents will require more than the normal letters, numbers or punctuation symbols visible on the keyboard. These include special symbols and characters such as Copyright © or Trademark ™. You may also need to change the currency symbol, for example to Euros €. All these symbols can be added from the **Insert** option.

insert symbols

1 Click on the page where you want the symbol to appear.

2 Click on the **Insert** tab.

3 On the far right of the ribbon, click on the **Symbol** button.

4 If appropriate, click on a recently used or common symbol showing here to add it to your text.

5 To find other symbols, click on **More Symbols**.

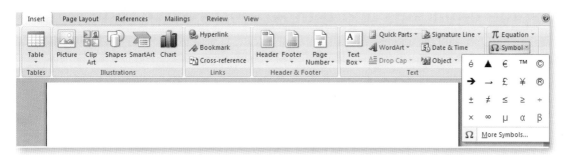

Fig. 1.99 Insert symbol

6 In the box that opens, recently used symbols will be readily available.

7 You may need to click in the **Font** and/or **Subset** box to choose from a different range of symbols but take care that, when inserted in the document, symbols are in the appropriate case and style.

8 Scroll down through the examples until you find the symbol you want.

9 Click on it and then click on the **Insert** button.

10 To insert **Special Characters**, find them by clicking on the tab.

11 Click on **Close** to leave the box and return to your document, or continue selecting and inserting more symbols.

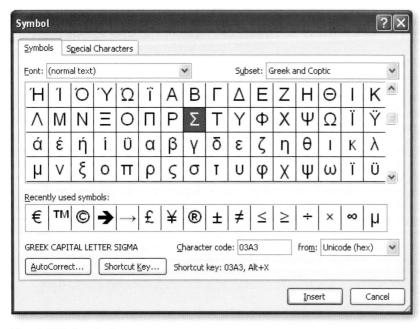

Fig. 1.100 Symbols

Note that although there are symbols available (for example, normal text – Latin–1 Supplement), to add European letters with accents such as acute, grave or umlaut, you can also use the keyboard.

type accented letters

1 Hold down **Ctrl**.

2 Press the key for any lower punctuation symbols such as ' or `

3 Hold **Shift** and then press for upper punctuation symbols such as ^ or :

4 Use:

- **Ctrl 'e** for é
- **Ctrl Shift ^ a** for â
- **Ctrl `** (next to 1 on the keyboard) **e** for è
- **Ctrl Shift :** (colon) **a** for ä

Typing equations and temperatures accurately

Instead of using symbols to type entries such as 35°C or C_6O_6 in a document, you format the characters so that they are smaller and sit above or below the normal typing line.

Characters above are **superscript** and characters below are **subscript**.

apply superscript or subscript

1 Type the letter or number as normal (for degrees of temperature, for example, use a lower case o).

2 Your entry might look like: **H2O** or **25oC**.

3 Select the relevant character with the mouse.

4 Click on the appropriate button in the **Font** group on the **Home** tab.

5 To continue typing normally if you have formatted the last character in this way, you will need to click the button again.

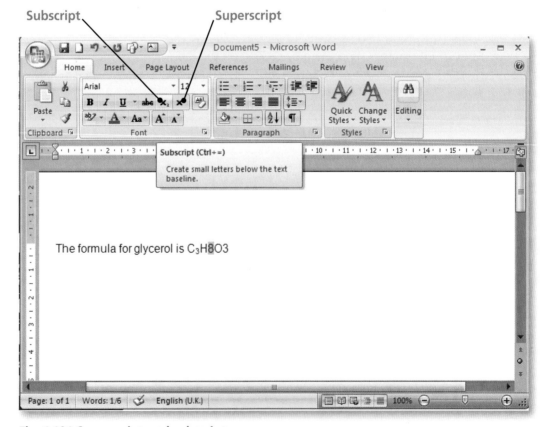

Fig. 1.101 Superscript and subscript

Automatic superscript

You are likely to find that, as you type the ordinals 1st, 2nd and so on, superscript is applied automatically and entries become 1st, 2nd, etc. This means that **AutoCorrect Options** have been applied. You can change the rules controlling such replacements by taking off ticks in the relevant checkboxes.

change AutoCorrect Options

1 Click on the **Office** button.

2 Select **Word Options**.

3 Click on **Proofing**.

4 Click on the **AutoCorrect Options** button.

5 On the **AutoFormat** tab, take off ticks in any boxes where you do not want the changes made automatically.

6 On the **AutoCorrect** tab you will find more rules and a list of replacements that you can delete or add.

7 Click on **OK** to confirm any new settings.

Click to prevent replacement

Fig. 1.102 AutoCorrect Options

Check your understanding 22

1 Open the file *Pharaohs* on the CD-ROM accompanying this book.

2 After the words *triangular tomb* in the third paragraph, insert a symbol to represent a pyramid or triangle (for example, from **Windings 3 Font** or **normal text – subset Box Drawing**).

3 At the end of the document, leave a clear line space and enter the following text, making sure the accents are included: **One famous French Egyptologist was Georges Aaron Bénédite (1857–1926) who investigated the Valley of the Kings.**

4 Start a new paragraph and type the following: **During the summer months the climate in the inland desert areas can vary widely from 7°C at night, to 43°C during the day.**

5 Save as *Pyramid*.

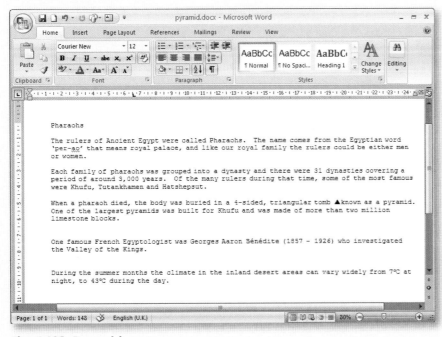

Fig. 1.103 Pyramid

Adding page numbers, date and time

It is very easy to add page numbers or the date or time to your documents. They are all available from the **Insert** tab.

number pages

1 Click on the **Insert** tab.
2 Click on **Page Number**.
3 In the gallery of styles you can choose a position and type of number.
4 Select from the **Page X of Y** section to include the number of actual pages in the document – this is useful when you want to indicate a document's length.

Format Page Numbers

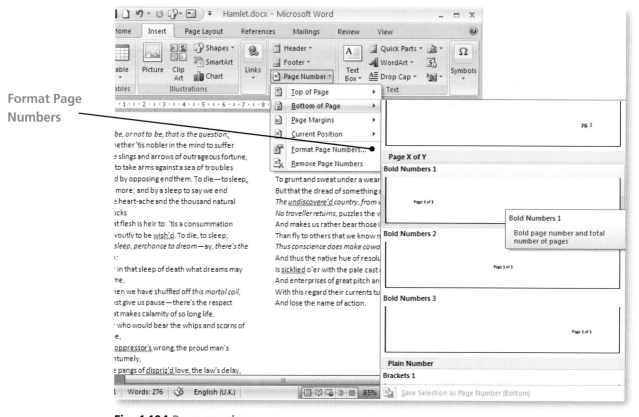

Fig. 1.104 Page numbers

Start number   Style

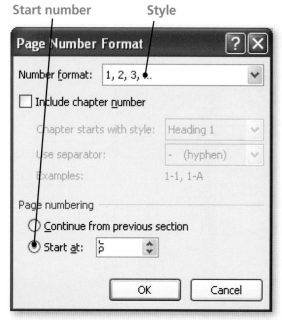

Fig. 1.105 Page format

5 Click on the **Format** option if you want to change:
   • which page number the document will start at
   • the style of numbers.

add the date or time

1 Click on the **Insert** tab and select **Date & Time**.
2 When the window opens, select your preferred example.
3 Make sure the Language is set to English UK to put the day before the month.
4 Click in the update checkbox if you are delaying printing but want the current date on any document. (Note that, if you do this, the document will not display the date it was created when you refer back.)

Headers and footers

Long documents often need a range of different information including the title or author added at the top or bottom of each page. These entries are known as **headers** (top) or **footers** (bottom). You can also add more specific information such as the file name or folder pathway. The value of using headers and footers is that they do not interfere with the main layout of the document pages. Also, entries such as date, time or page numbers inserted from the menu can be set to update automatically.

To add several items, move across the header or footer box by pressing the **Tab** key or double clicking the mouse, and move onto a new line by pressing **Enter**.

add headers and footers

1 Click on **Header** and select a style such as **Blank**.

2 When dotted lines show the position for the header box, start making entries.

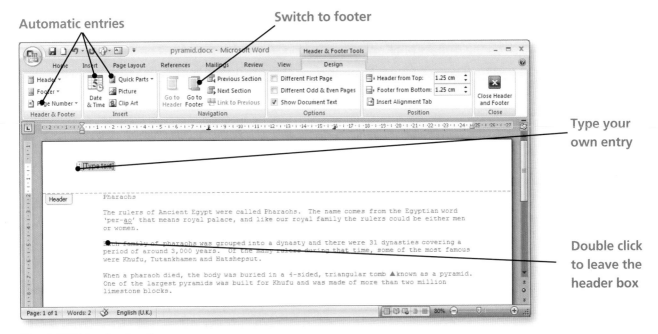

Fig. 1.106 Header toolbar

3 Click on buttons on the **Header and Footer Tools – Design** tab to add extra fields.

- **Page Number** offers styles to choose from.
- **Date & Time** will open the dialog box where you can choose how the date/time entry will be formatted. Click on the checkbox to update the entry automatically.
- **Quick Parts – Field** will open the list of fields. Scroll down and click on **FileName** to add this. Click on the checkbox if you also want to show the folder pathway and click on **OK** to return to the document.

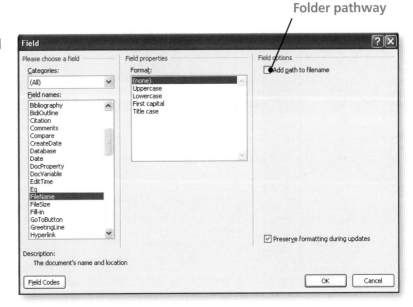

Fig. 1.107 Field in header

4 To make entries in the footer, click on the **Go to** button.

5 To return to normal typing, double click on the greyed out document text or click on the **Close Header and Footer** button.

6 Reopen the header or footer box by double clicking on an entry.

Bullets and numbering

Lists can be enhanced and the contents are often displayed more clearly if **numbers** or **bullets** are added. The style of number or bullet can be changed by selecting a different option, and it is easy to remove them if they are unwanted.

If you apply bullets or numbering before typing, these will be added each time you press **Enter**.

When bullets or numbers are added, the text will be indented automatically and you can change the distance that both the bullets and the text are from the left margin.

add bullets or numbers

1 Select list items or follow these steps first.

2 Click on the **Bullets** or **Numbering** button. The default style will be applied.

3 To change this, click on the drop-down arrow next to the button and select an alternative.

4 Take off bullets or numbering by clicking on **None** or selecting the list and clicking off the button.

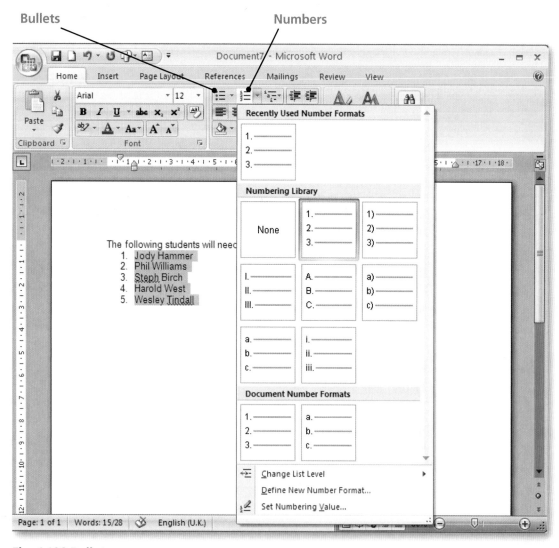

Fig. 1.108 Bullets

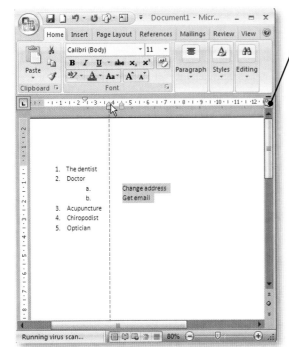

View ruler

Fig. 1.109 Number ruler position

5 If text or a bullet or number is in the wrong position on the page, select the list item and then drag the relevant marker across the ruler. (Click on the **View Rulers** button at the top of the vertical scroll bar if it is not displayed.)

or

6 Right click the selected list items and select **Adjust List Indents**. You can then change the measurements in the various boxes.

Fig. 1.110 List indent box

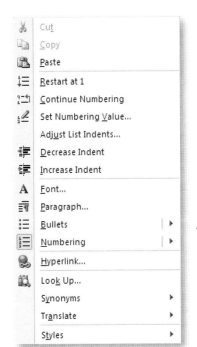

Fig. 1.111 Change numbers

7 To change the numbering, for example if a list is split across an image, right click on one of the numbers and select the appropriate menu option:
- Restart at 1
- Continue from an earlier number
- Change the start number (**Set Numbering Value**).

Check your understanding 23

1 Start a new document.

2 Type the following text exactly as set out below:

Countries I have visited

Over the past few years I have visited a number of different countries and these include:

Sweden
India
Mexico
Germany
Ireland

My favourite place was Stockholm.

3 Save as *Countries*.

4 Apply bullets to the list of countries.

5 Change to a different style of bullet.

6 Double space the list.

7 Add a footer showing the page number and file name.

8 Close and reopen the file.

9 Add the word *Geography* as a header.

10 Change to landscape orientation.

11 Save and close the file.

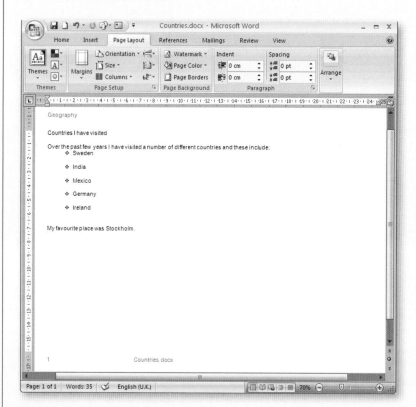

Fig. 1.112 Countries

CLAiT Assignment

For this assignment, a friend or colleague will need to send you an email entitled **Next play** about future performances by the company's drama group, attaching the archive **Plays** provided. It should just say the files you wanted are attached.

Task 1

1   Enter details of the following two contacts into your address book:

NAME	JOB TITLE	HOME PHONE	EMAIL ADDRESS
Sarah Khan	Marketing Assistant	01775 4993426	s.khan2@pearson.com
Daisy Wells	Telesales Manager	01775 3882399	daisy.wells@pearson.com

2   Create a distribution list named **Drama group**.

3   Add the two new contacts to the list and then save and close.

4   Print a copy of the contacts showing their full details.

5   Print a copy of the distribution list showing its name and full details of the members.

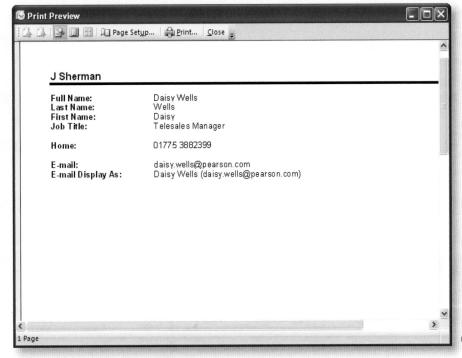

Contact details – one example

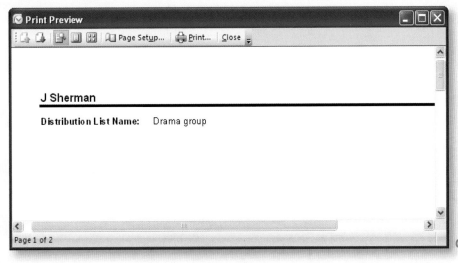

Contacts1

Contacts2

Task 2

1 Create a new folder inside your Inbox labelled **Drama**.

2 Read the message sent to you entitled **Next play**.

3 Store the archive/zipped attachment named **Plays** outside your mail system.

4 Add the email address of the sender to your address book.

5 Move the message **Next play** to the folder **Drama**.

6 Locate the zipped archive **Plays** and extract the 3 files it contains: **Death on the Nile**, **Audiences and Hamlet**. Store these in the same location as the archive.

Task 3

1 Create an email signature named **Artistic** that has the following details:

   Your full name

   Artistic Director

   Pearson Players

2 Take a screen print to show the signature.

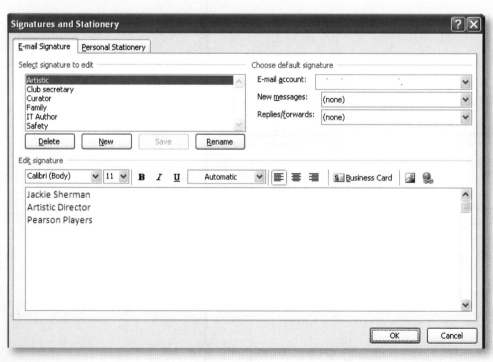

Signature

3　Compose a new email message.

4　Address it to the **Drama group** whose details you should retrieve from your contacts.

5　Enter the following details:

Subject: **Our next performance**

Main message: **As you know, we will be performing for the Board in six months. Let me know which play we should tackle by the end of the week. I will sort out rehearsal rooms and refreshments.**

6　Add the signature **Artistic** you created earlier.

7　Attach the three files you extracted from the **Plays** archive as separate attachments. Do not attach the archive.

8　Copy the message to **room_bookings@pearson.com**

9　Send a blind/confidential copy to your friend or colleague, retrieving their address from your contacts.

10　Mark the message high priority.

11　Before sending, take a screen print to show all the above features.

12　Send the message, ensuring that a copy is saved.

13　Print a copy of the sent message.

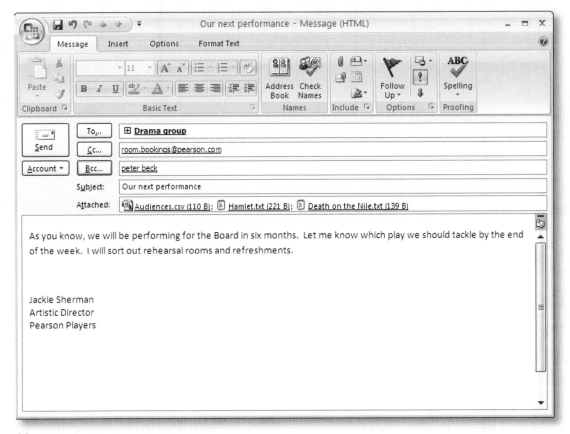

Message to group

Task 4

1　Use your calendar to schedule the following meetings and appointments for next week:

DAY	Start time	Duration	Location	Topic	Recurring	Alarm
Wednesday	10.30am	1 hour	My office	Finance	Weekly	Yes
Wednesday	12.30pm	1 hour	Boardroom	Lunch	Monthly	Yes
Thursday	5.00pm	2 hours	Room 702	Drama rehearsal	No	No

2　Print a copy of the diary pages for these days only showing the above details.

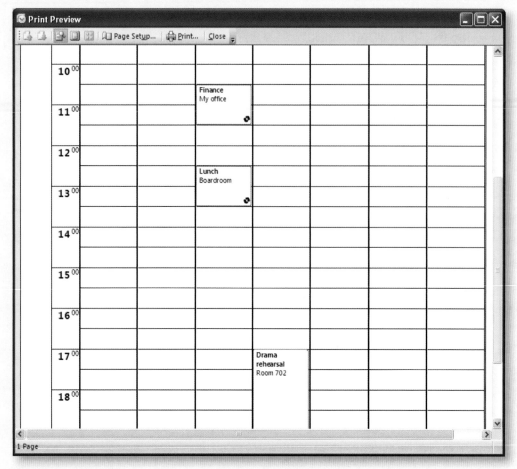

Details on calendar

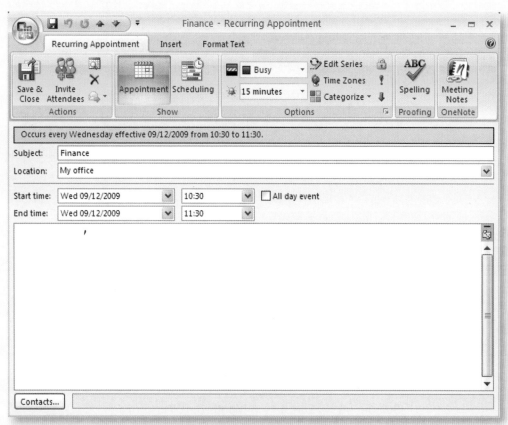

Alarm set

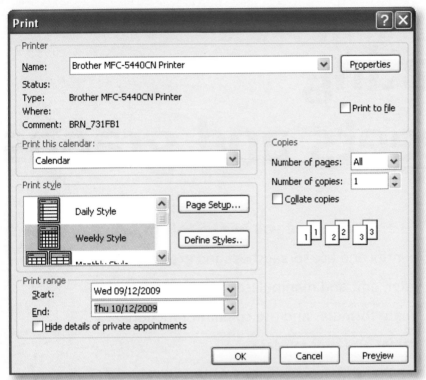

Print calendar pages

3   Open a copy of the file
    Death on the Nile
    you extracted and saved.

4   Copy the contents into a
    Notes page.

5   Print a copy of the Notes page.

> 1. Take up Deirdre's dad's
> offer to build the Nile
> scenery
> 2. Large cast list – must
> contact Smithfield & Co for
> combined performances

Death on the Nile notes

Task 5

1   Take a screen print showing the contents of the ***Drama*** folder.

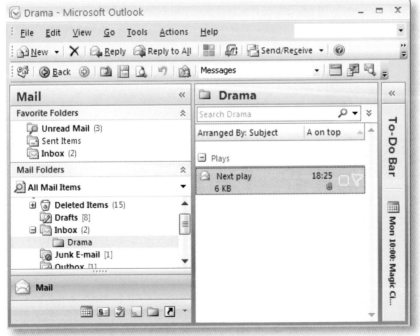

Drama folder

2   Make sure all the contents are clearly shown.

3   Save and close your email software.

Manipulating spreadsheets and graphs

After completing this unit, you will have a more indepth understanding of the spreadsheet application Excel 2007. You will be able to design and create spreadsheets and use complex formulas and functions. You will also be able to analyse and interpret data to produce a variety of charts and graphs as well as import data from elsewhere.

At the end of the unit you will be able to:

➡ identify and use spreadsheet and graph software correctly

➡ enter, edit and manipulate data

➡ create formulas and use common functions

➡ format and present data

➡ link live data from one spreadsheet to another

➡ import data files

➡ present data using charts and graphs and format these appropriately

➡ use graphs to extrapolate information to predict future values

➡ use spreadsheets to solve problems and project results.

Creating and formatting spreadsheets

Once you have launched the spreadsheet program, a workbook will open and you can start entering and formatting your data. You will find all the basic formatting tools on the **Home** tab, or you can open the relevant dialog boxes to make more complex changes.

With large spreadsheets, it can take time to format a number of cells in the same way. It is quicker to select all the cells first so that you can apply the same formatting to them at once.

Select all

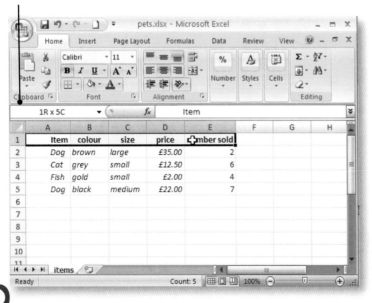

select cells

- Click on a single cell.
- To select adjacent cells, click and drag across a range of cells when the pointer shows a white cross.
- Select non-adjacent cells by selecting the first range of cells and then holding **Ctrl** as you select cells in different rows or columns.
- Click on the **Select All** button in the top corner of the spreadsheet to select all the cells on the sheet.

Fig. 2.1 Select cells

Having formatted a range of cells, you may find you want to apply the same formatting elsewhere. Do this by using **Format Painter** to *paint* the new formatting over other cells.

use Format Painter

1 Select a cell that has been formatted.

2 Click on the **Format Painter** button to format a single cell or cell range. Double click on the button if you want to apply the formatting several times, as this keeps the button turned on.

3 The selected cell(s) will now display a flashing, dotted border so that it is easy to see the formatting you will be applying.

4 Select the cells to receive the new formatting. The pointer will show a brush.

Format painter button

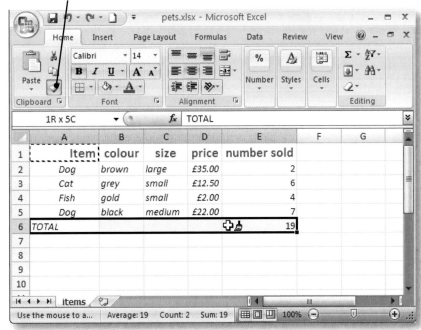

Fig. 2.2 Format Painter

5 If necessary, repeat this several times and then click off the **Format Painter** button.

Formatting text

You can select options on the **Home** tab in the **Font** section to:

- select a different type of font
- set a new font size – if the size is not available from the drop-down arrow in the box, type in your own figure and then press **Enter** to confirm the setting
- increase or decrease the font size in steps using the **Increase** or **Decrease font** buttons
- apply a colour
- add emphasis such as bold, italic or underline

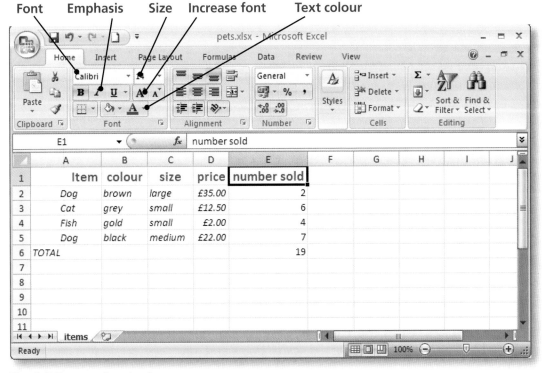

Fig. 2.3 Formatting tools

Check your understanding 1

1 Open the file *Property* provided on the CD-ROM accompanying this book.

2 Format the title to bold, Times New Roman size 14, green.

3 Format the column headings to bold, font size 12

4 Format the data to font size 11.5.

5 Format the property types to italic, red.

6 Save as *Rent* and then close the file.

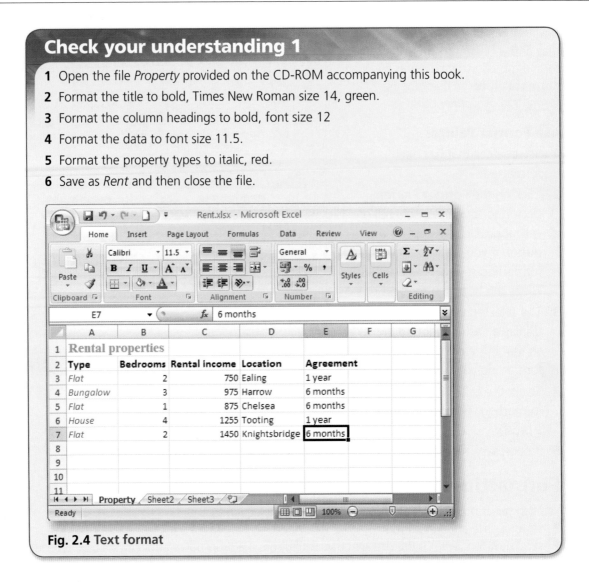

Fig. 2.4 Text format

Formatting numerical data

Although the underlying values will remain the same, you can display figures in a range of different formats. These include:

- changing the number of visible figures after the decimal
- currency symbols
- showing a percentage
- having commas separating the thousands
- selecting a different time or date style.

Sometimes an automatic format is applied incorrectly and you will need to reformat. For example, dates can appear as a string of numbers.

To show trailing 0s, apply a Text format. To start a number with zero, use the Text format or type an apostrophe into the cell before entering the number.

Numerical data is aligned automatically on the right of the cell. If it appears on the left, you may not be able to format the numbers correctly. This can happen, for example, if you include letters as well as figures or keyboard strokes such as a space.

In some cases, the column may not be wide enough to display all the figures. When this happens, you will see a row of #####. You need to widen the column to display the entry in full.

Some formats can be found on the **Home** tab but you can make more detailed changes from options on the **Number** dialog box.

format numbers using the toolbar

1 Select the entries and click on the appropriate toolbar button.

2 Click on the drop-down arrow in the **Number Format** box for a further range of options.

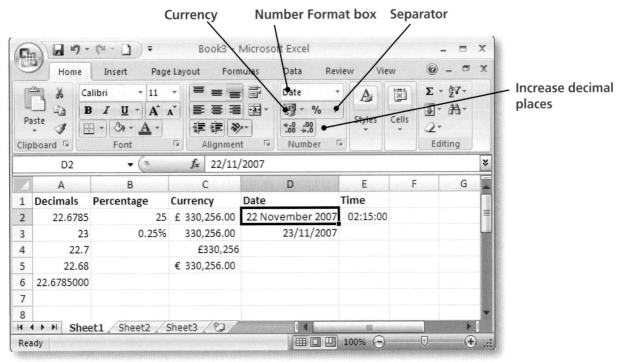

Fig. 2.5 Number formats on tab

format numbers from the menu

1 Select the cell(s) to format and then click on **More Number Formats** or the arrow in the **Number** group on the ribbon to open the dialog box.

2 You can also right click on a cell and select **Format Cells** to open the box.

3 On the **Number** tab, click on the appropriate category such as date or currency. The **General** category will display the entry exactly as you type it.

4 If relevant, select the number of decimals to display and add or remove the comma separator. The term for whole numbers is **Integer** format.

5 For a currency symbol, select the preferred symbol or click on **None** to remove one.

6 For dates, select a style such as **short** (12/2/08) or **long** (12 February 2008).

7 Check the preview and then click on **OK** to accept the formatting.

Note that you must always select a numerical entry as the first cell in a range, rather than a column or row heading, or the preview will not show the formatting to be applied.

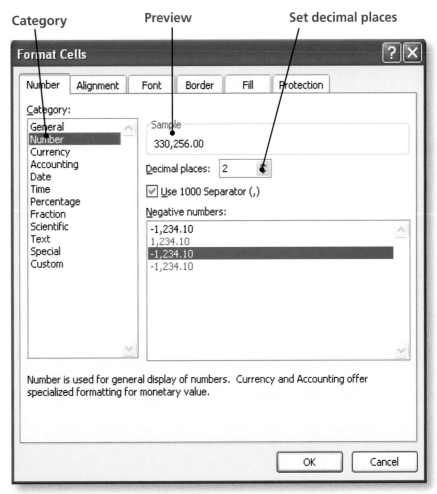

Fig. 2.6 Number format dialog box

Check your understanding 2

1 Start a new workbook.

2 Enter the following across row 1:

    a 2/5/07 in A1

    b 22.675 in B1

    c 0.234 in C1

    d 4.5 in D1

3 Format the date in A1 to a long date.

4 Format the entry in B1 to integer format (no decimals).

5 Format the entry in C1 as a percentage.

6 Format the entry in D1 as currency with 2 decimal places.

7 Close the file without saving.

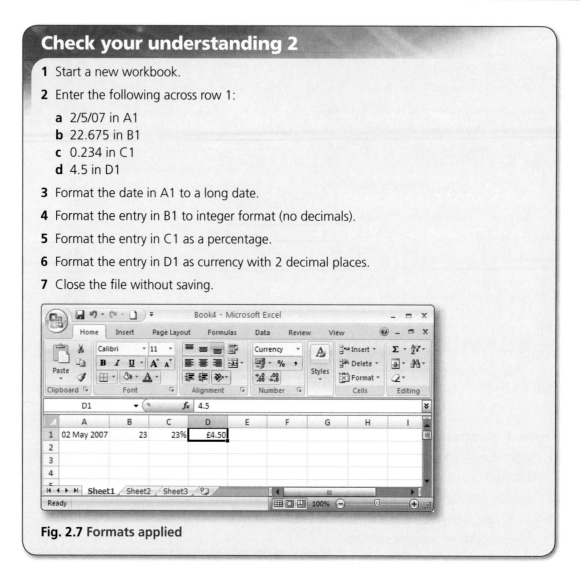

Fig. 2.7 Formats applied

Text orientation and alignment

With large spreadsheets, you may find that column or row headings take up too much space. Instead, you can angle text or even position it vertically in the cell. You can also set the text alignment horizontally across the cell or vertically at the top, centre or bottom.

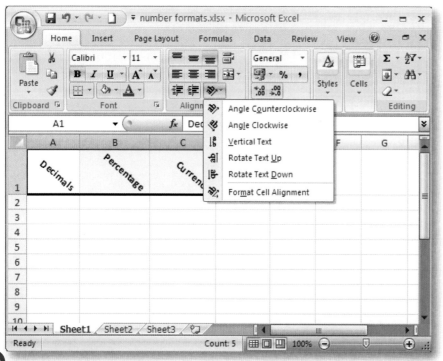

set text orientation

1 Select the text to be amended.

2 Click on the drop-down arrow in the **Orientation** box and select an alternative.

3 To set text at an exact angle, click on **Format Cell Alignment** to open the dialog box.

Fig. 2.8 Text orientation

4 On the **Alignment** tab, drag the red pointer to a new position in the **Orientation** box to set the angle by eye.

5 You can also click on the up or down arrows in the **Degrees** box. (A plus figure angles the text up to the right and a minus figure angles the text down to the right.)

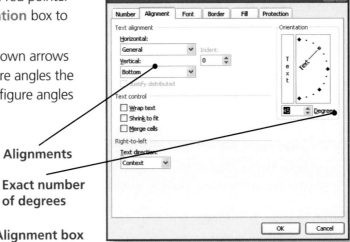

Alignments

Exact number of degrees

Fig. 2.9 Alignment box

set horizontal or vertical alignment

1 Select the cells.

2 Choose an alignment from the **Home** tab. There are three vertical and three horizontal alignments to select from and you can set both horizontal and vertical alignments at the same time.

3 For other options, open the **Format** dialog box and click on the **Alignment** tab.

4 Click in any of the alignment boxes for options such as **Justify** or **Distributed**.

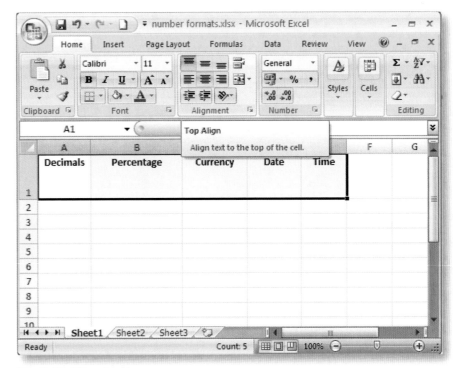

Fig. 2.10 Align buttons

Wrapping cell contents

For a very long entry that would take up too much space if displayed on a single line, you can wrap the text in the cell.

wrap text in a cell

1 Select the cell(s).

2 Click on the **Wrap Text** button on the **Home** tab or select this option on the **Alignment** tab in the **Format Cells** dialog box.

3 Widen the column by eye if you want the text to take up fewer lines in the cell.

4 Drag the lower boundary down if you want a very thin column.

Wrap button

Fig. 2.11 Wrap text

Merging cells

For a spreadsheet displaying data in a number of columns, it is more attractive and easier to work with if the appropriate heading is centred above them. To do this, you need to merge the cells so that the heading can be centred in a single, large cell stretching across the full width of the spreadsheet data.

merge cells

1 Select all the cells to be merged.

2 Click on the **Merge and centre** button on the toolbar or select this option on the **Alignment** tab in the **Format Cells** dialog box.

3 The cell will now be a single, wide cell and the text will be centred inside it.

Cells A1 and B1 seleted   **Merge and centre**

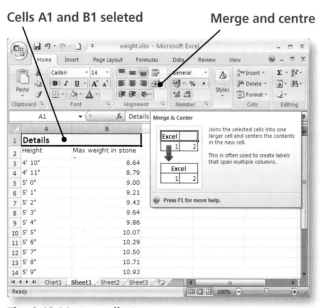

Fig. 2.12 Merge cells 1

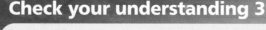

Fig. 2.13 Merge cells 2

Check your understanding 3

1 Open the file *Gardening* provided on the CD-ROM accompanying this book.

2 Centre the title across the width of the data.

3 Reorientate the names of the products to read from bottom left to top right at 25 degrees.

4 Wrap the entry in A2 to keep the column narrow but display all the text.

5 Align all the column headings except *Products sold* in the centre right of the cells.

6 Format *Unit prices* and *Final prices* to Currency with 2 decimal places.

7 Format the title to bold, font size 16.

8 Format the column headings to bold, font size 13.

9 Save as *Gardening2* and close the file.

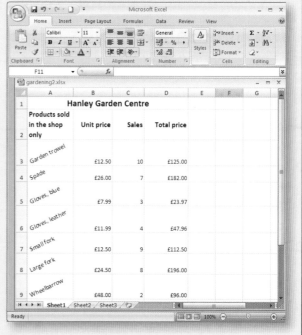

Fig. 2.14 Gardening2

Borders and shading

One way to emphasise a particular section of a spreadsheet is to add borders or shade the cell background. These options are available on the **Home** tab or you can open the **Format Cells** dialog box for further options.

border cells

1 Select the cell range.
2 Click on the drop-down arrow next to the **Borders** button and select a style of border.
3 You can also add a line colour and change the line style.
4 For further options, click on **More Borders**.
5 On the **Borders** tab, click on one of the **Preview** buttons to take off sections of a border such as the top or side, and choose different line styles and colours from the drop-down options.
6 Click on **None** to remove a border.
7 Click on **OK** to confirm the settings.

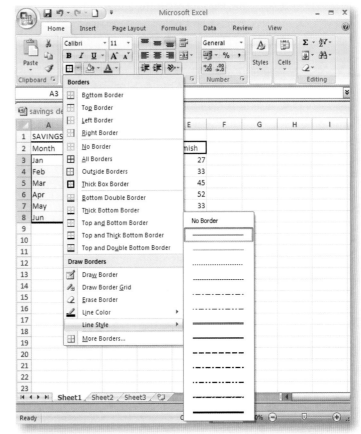

Fig. 2.15 Borders

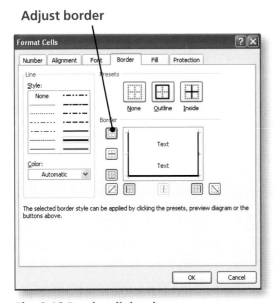

Adjust border

Fig. 2.16 Border dialog box

add shading

1 Click on the **Fill** button on the **Home** tab for a range of colours.
2 If necessary, change the font colour so that entries remain visible.
3 Click on **More Colours** to open the **Format Cells** dialog box.
4 On the **Fill** tab you can select plain colours and patterns or click on **Fill Effects** to mix two colours together as a gradient.
5 Remove shading by clicking on **No Fill**.

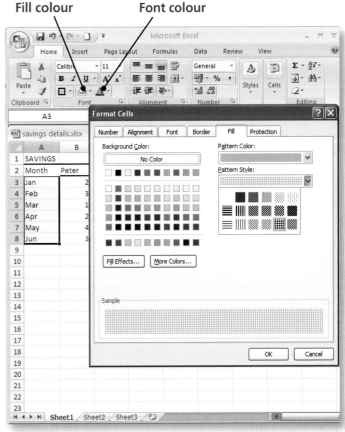

Fig. 2.17 Shading

Check your understanding 4

1 Open the file *Savings details* provided on the CD-ROM accompanying this book.

2 Centre the heading across the data columns and emphasise the text.

3 Format the numerical data to currency with no decimal places.

4 Shade the cells containing names dark red and apply a lighter font colour if necessary to make sure the details are still visible.

5 Border all the currency entries with a thick single line border.

6 Shade the months green (but *not* the column heading).

7 Save as *Savings colour*.

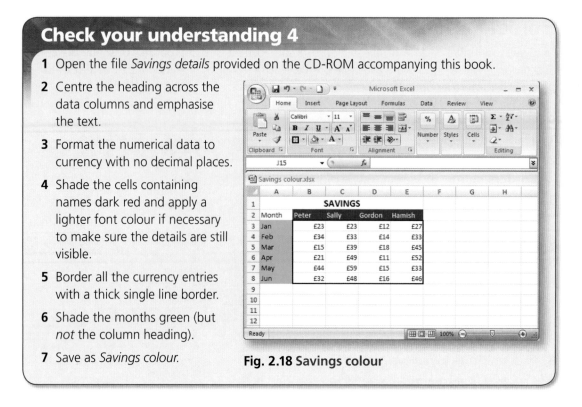

Fig. 2.18 Savings colour

Editing data

As you Learned at level 1, when you want to make changes to data entered into a spreadsheet, you can move or copy it by dragging, or use cut, copy and paste. You can also replicate entries using the AutoFill functions in Excel.

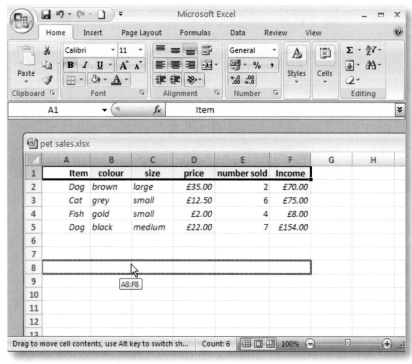

Fig. 2.19 Move by dragging

move data by dragging

1 Select the cell(s) and then reposition the pointer.

2 As you bring it up to the edge of a selected cell or cell range, it will display a white arrow with four black arrows at the tip.

3 Hold down the mouse button and drag the cells to their new position.

4 A dotted outline will show where the cells will move to.

5 Let go of the mouse when the cells are in position.

copy cells by dragging

1 Repeat the above but hold down **Ctrl** as you drag.

2 You will see a plus sign at the end of the arrow.

If you arrange two workbooks side by side on screen, you will find you can still use dragging to move or copy cells.

arrange windows

1 Click on the **View** tab and select **Arrange All**.

2 Click on an arrangement such as vertical or horizontal tiling.

3 All open spreadsheets will be visible at the same time.

move data using Cut and Paste

1 Select the cells.

2 Right click and select **Cut**, or find this option on the **Home** tab.

3 Click on only the *first* cell to take the entry. This can be on the same or a different sheet, or in a new workbook.

4 Right click and select **Paste**.

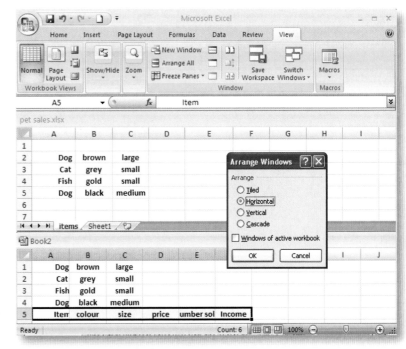

Fig. 2.20 Arrange windows

copy data using Copy and Paste

1 Select the cells.

2 Right click and select **Copy**.

3 Click on the first cell to take the copied entries.

4 Right click and select **Paste**.

5 Dotted flashing lines will mark the copied entry so press the **Escape** key to remove them.

Copy   Cut   Paste

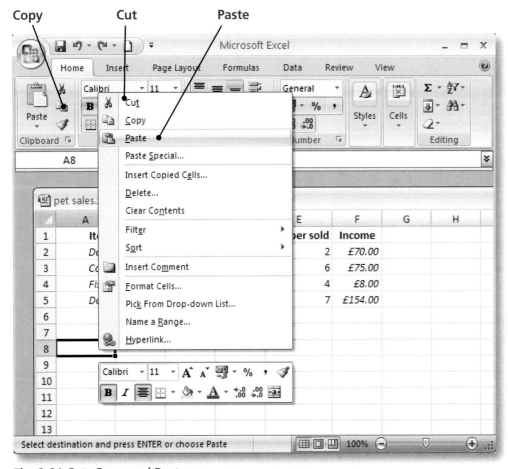

Fig. 2.21 Cut, Copy and Paste

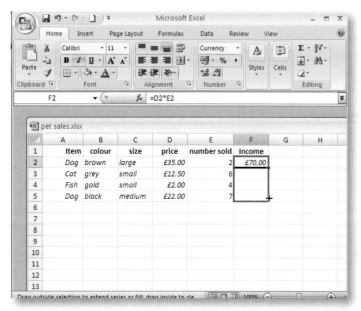

Fig. 2.22 Fill handle

Fill button

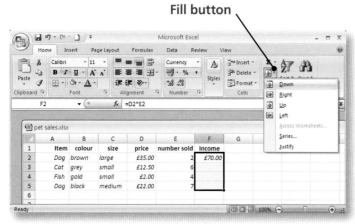

Fig. 2.23 Fill button

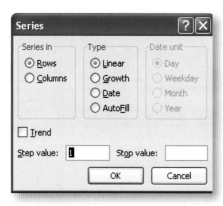

Fig. 2.26 Series menu

replicate entries using the fill handle

1 Select the first cell.

2 Move the pointer to the fill handle in the bottom right-hand corner of the cell.

3 Drag the pointer down a column or across a row when the pointer shows a small black cross.

- Entries such as text or numbers will copy exactly.

- Formulas will replicate using relative cell addresses, i.e. =D2*E2 will become =D3*E3, =D4*E4, etc.

- Dates will replicate incrementally, i.e. January will become February, March, April, etc.

- To copy an incremental series, or stop the automatic increment, select two cells and then copy down from the *second* one.

Take care not to 'overshoot' when replicating formulas as this will result in adding data to what should be blank cells.

replicate entries using the menu

1 Select the cell containing the entry to be copied as well as all the cells to receive the copy.

2 On the **Home** tab, click on the drop-down arrow in the **Fill** box and select the direction for the copies.

3 All the cells will be filled.

4 To set the steps and type for an incremental series, click on **Series** and select the appropriate options. For example, you could copy the series 20, 30, 40 by selecting **Linear** and setting the step value as 10, or you could copy a series displaying only Saturdays in May by selecting **Date and Day**, setting a step value of 7 and a stop value of 31/5/09.

Using special paste options

Where the data being copied or moved involves formulas that are referencing particular cells on a spreadsheet, you may find pasting results in errors. There is therefore a facility to use a special paste function. You can also use this option to copy formatting to different parts of a worksheet.

paste selectively

1 Select the cells to be copied or moved.

2 Click on **Copy/Cut**.

3 Click on a cell on the worksheet to receive the entries or select a range of cells to receive copied formatting.

4   Click on the drop-down arrow under the **Paste** button on the **Home** tab.

5   Select values or formulas.

6   If you click on **Paste Special** you open the menu that offers more options, including pasting only the formatting.

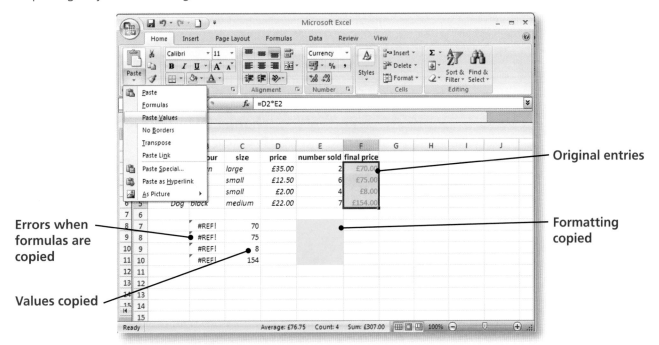

Fig. 2.24 Paste functions

Check your understanding 5

1   Open the file *Photos* provided on the CD-ROM accompanying this book.

2   Move the heading to a new worksheet.

3   Copy the entries in rows 3–7 to rows 10–14.

4   Copy only the values in cells G3–G7 to cells A1–A6 on the worksheet containing the heading.

5   Format the heading to italic, font size 14, dark blue.

6   Copy this formatting to all the cells on Sheet 1 containing data.

7   Widen columns and wrap text to improve the clarity of the data on Sheet 1.

8   If necessary, reapply date and currency formatting.

9   Save as *Photos copied* and close the file.

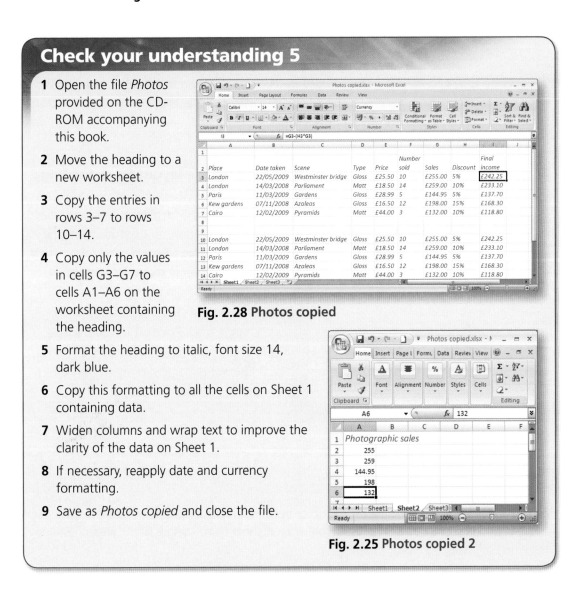

Fig. 2.28 Photos copied

Fig. 2.25 Photos copied 2

Spreadsheet layout

The choices that you can make when setting out a spreadsheet before printing include changes to the:

* margins
* orientation
* paper size
* headers and footers.

These options are all available from the **Page Setup** dialog box or buttons on the **Page Layout** or **Insert** tabs, but to see the full effect it can be helpful to be in **Print Preview**.

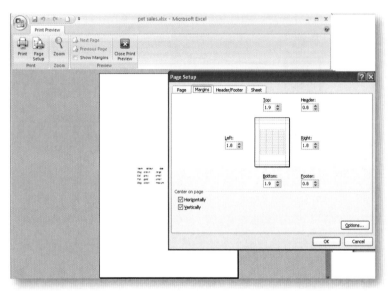

Fig. 2.26 Margins

change layout options in Print Preview

1 View the spreadsheet by going to **Office Button – Print – Print Preview**.

2 Click on **Page Setup**.

3 On the **Margins** tab, increase or decrease individual margins or click to centre the data on the page.

4 On the **Page** tab, click to change orientation or select a different size paper to print on.

5 Click on **Headers/Footers** to add entries at the top or bottom of the worksheet.

6 If you click on the **Show Margins** button on the **Print Preview** tab, you can also drag the margins to position them by eye.

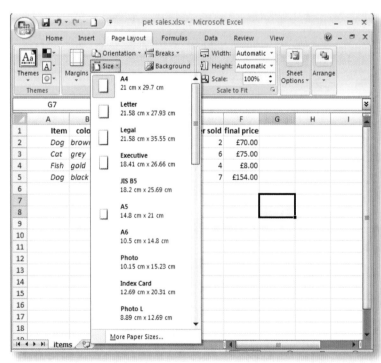

Fig. 2.27 Paper size

change page layout from the Page Layout tab

1 Click on the **Margins** button and select a layout or click on **Custom Margins** to open the **Page Setup** dialog box where you can change measurements in the different margin boxes.

2 Click on the **Orientation** button to change from **Portrait** to **Landscape**.

3 Click on the **Size** button to select a different paper size or open the dialog box for further choices.

Headers and footers

You can add a range of basic data to the top or bottom margins of a spreadsheet including page numbers, the date or a title as well as the file or worksheet name, details of the creator and any file (folder) pathway.

add entries in Headers and Footers

1 On the **Insert** tab, click on **Header or Footer**.

2 This adds a box at the top or bottom of the sheet and you can type in an entry directly.

3 Click in the box and then click on any of the buttons on the **Header and Footer Tools Design** tab to add page numbers, dates, file name, etc.

4 If you click to the right or left of the box, you can add entries in other parts of the **Header or Footer** area.

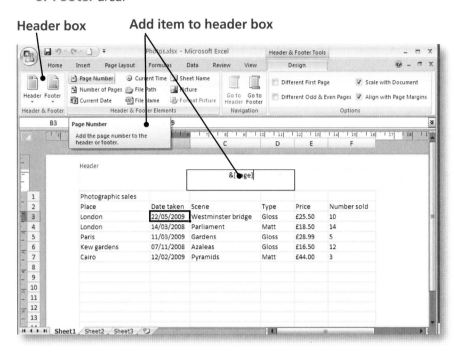

Header box

Add item to header box

Fig. 2.28 Header on sheet

5 Click on the drop-down arrow below the **Header or Footer** button for a list of pre-set entries to add.

6 Return to the header or footer box to amend or delete entries by clicking on the **Header or Footer** button.

add Headers and Footers using the menu

1 Open the **Page Setup** dialog box from the **Page Layout** tab or from **Print Preview**.

2 Click on the **Header/Footer** tab and either select an automatic entry or click on **Custom** to create your own.

3 Hover the mouse over any button to see what entry it will add.

4 Position the cursor to the left, centre or right before adding entries from the toolbar buttons or by typing them in, to place the entry in that section of the spreadsheet.

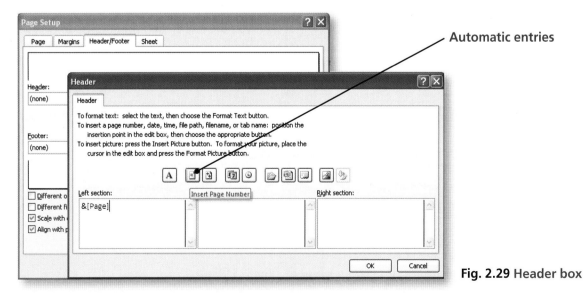

Automatic entries

Fig. 2.29 Header box

Check your understanding 6

1 Open the file *Orders* provided on the CD-ROM accompanying this book.

2 Change the left and right margins to 3cm.

3 Change to landscape orientation.

4 Add the following in the header area:

  a page number and number of pages – include extra text so that it is displayed as 'page number' out of 'number of pages'

  b date

  c time

5 Add your name and the worksheet name as a footer.

6 Now change the left margin to 4cm and the top margin to 10cm.

7 Remove the date entry in the header.

8 Save as *Orders layout* and close the file.

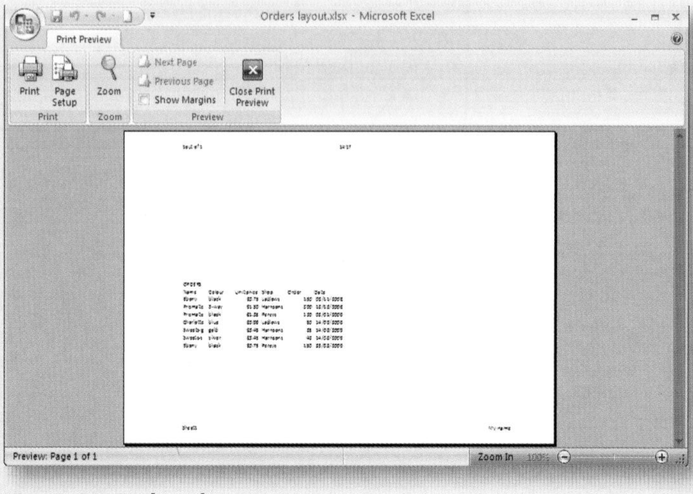

Fig. 2.30 Orders layout

Find and replace

As with all Microsoft Office applications, you can use the search and replace tools to locate or amend data quickly.

In Excel 2007, you can search for data:

- on one sheet or throughout the workbook
- by rows or columns
- that is a value or is within a formula
- where particular formatting has been applied
- that is case sensitive
- that is part of or an entire cell entry.

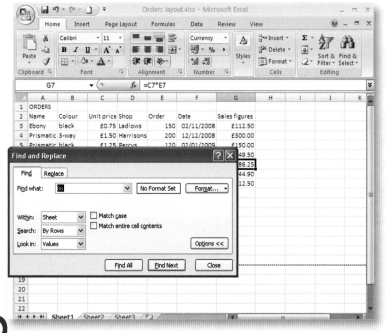

locate data using Find

1 Go to **Home – Find**.

2 In the box that opens, enter the data you are searching for in the **Find what:** box.

3 Click on **Options** and select from the drop-down lists or tick any checkboxes to set particular conditions for the search.

4 Click on **Find Next** to locate the first matching entry and continue clicking to find further matches.

Fig. 2.31 Find

replace data

1 Click on **Home – Replace** or click on the **Replace** tab in the **Find and Replace** window.

2 Enter the data you want to replace in the **Find what:** box and the replacement data in the **Replace with:** box.

3 Set any criteria as explained above.

4 To replace all entries, click on **Replace All**.

5 To first view all entries to be replaced, click on **Find All** and they will be displayed below the window.

6 To check each entry by eye, click on **Find Next**. To replace the entry click on **Replace**, and to skip and move to the next matching entry click on **Find Next**.

Check your understanding 7

1 Open the file *Numbers* provided on the CD-ROM accompanying this book.

2 Replace any entries for the whole number 200 with the number 20,000.

3 Make sure that entries in B1 and B28 have *not* been amended.

4 Close the file without saving the changes.

Freezing windows

When working low down on a large spreadsheet, it is easy to forget which column to enter your data into as the column headings will no longer be visible. In the same way, after scrolling across a wide spreadsheet you will no longer see row headings.

To prevent errors, you can freeze the column or row headings, or both sets of headings. They will now always be visible on screen. As you scroll down or across your data, earlier rows or columns will temporarily wrap underneath the headings.

freeze panes

1 Click on the **View** tab.

2 Click on **Freeze Panes** and then select which row and/or column to freeze.

3 A dark line will appear across the spreadsheet and headings will now stay visible on screen.

4 To remove the freeze and allow you to scroll normally, click on the **Unfreeze Panes** option.

An alternative method for keeping entries in view is to split the window. In this case, you can scroll in both windows.

Top row frozen

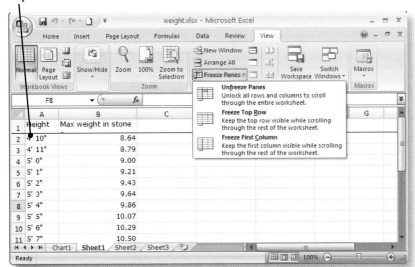

Fig. 2.32 Freeze

split windows

1 Click on the cell below or to the right of where you want the split to be created.

2 Click on the **Split** command on the **View** tab.

3 When you want to return to a single window, click on the **Split** command again.

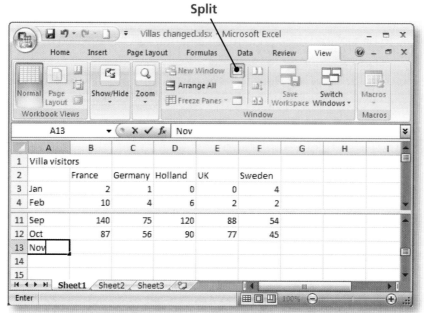

Fig. 2.33 Split

Inserting and deleting columns and rows

At Level 1, you learned how to add and remove extra columns and rows using options on the **Home** tab.

insert columns and rows

1 Click on the column or row marker or click anywhere in the column to the right or row above where the new one should appear.

2 Select more than one column or row for the number of new insertions.

3 Right click the mouse and select **Insert**. Select the correct option in the window that opens.

or

4 Click on the **Insert** button on the **Home** tab or click on the drop-down arrow for further options.

5 New columns or rows will appear and column letters and row numbers will be adjusted automatically.

or

6 Select the column to the left or row below, hold **Ctrl** and **Shift** and press + (plus).

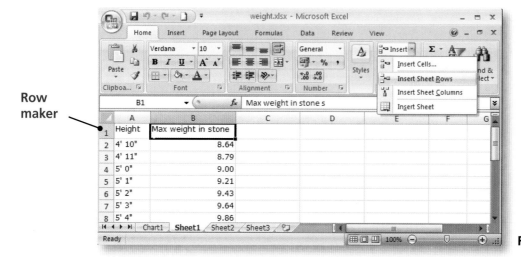

Fig. 2.34 Insert

delete columns and rows

1 Select the column, row or individual cells.

2 Right click and select **Delete** and choose the appropriate option.

or

3 Click on the **Delete** button on the **Home** tab.

4 If it is not clear which option should be selected, you need to click on the drop-down arrow in the box and select from the list.

or

5 Select the column or row, hold **Ctrl** and press – (minus).

There are various other methods you can use to remove data from a spreadsheet. These include:

- hiding columns or rows temporarily
- clearing the contents of cells.

hide columns or rows

1 Select the columns/rows.

2 Click on the drop-down arrow in the **Format** box on the **Home** tab.

3 Select an appropriate **Hide** option from the **Visibility – Hide & Unhide** menu.

4 To make the data visible again, open the menu and select **Unhide**.

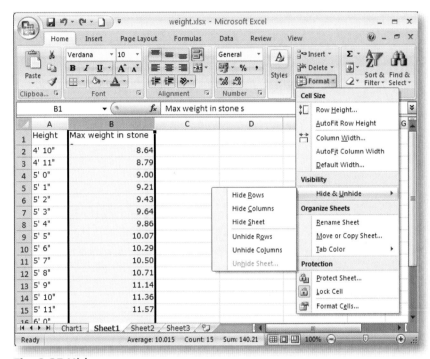

Fig. 2.35 Hide

clear cell contents

1 Select the data.

2 Press the **Delete** key.

or

3 Right click and select **Clear Contents**.

or

4 Click on the drop-down arrow next to the **Clear** button on the **Home** tab.

5 With this option, you can selectively clear the formatting, contents or both.

Clear button

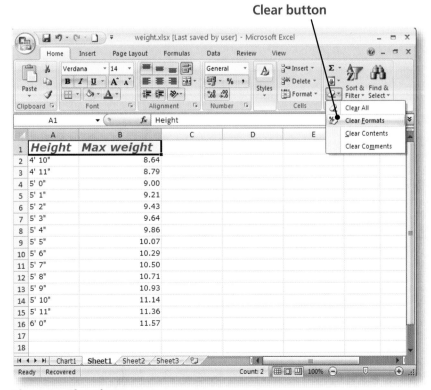

Fig. 2.36 Clear button

Relative and absolute cell references

As you learned in Level 1, formulas involve adding, subtracting, multiplying or dividing actual figures or the contents of cells. The cells in a formula are identified by their column header letter and row header number. This position of a cell on a worksheet is known as its **cell reference**.

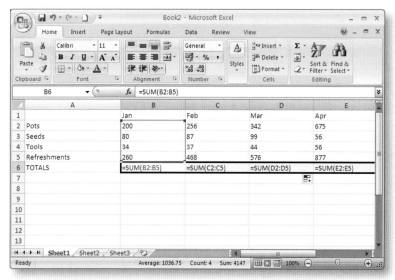

Fig. 2.37 Relative cell references

When you replicate a formula down a column or across a row, Excel makes use of **relative cell references**. This means that, when a formula totalling the contents of cells in column B is copied across the row, the replicated formulas will now refer to cells in column C, column D, column E and so on.

In the same way, when you copy formulas down a column, they will refer in turn to cells in Row 1, Row 2, Row 3 and so on.

It is quite common to want to refer to the same individual cell within a formula that is going to be replicated. For example, you may be replicating formulas that refer to a discount, unit of measure or converted currency entered into a single cell on the worksheet. Here, the cell reference must not change as you copy formulas down columns or across rows.

You fix the cell position by using its **absolute cell reference**. If the row stays the same but you want to change relative column positions, or vice versa, you can also set a **mixed cell reference**.

In order to replicate such formulas accurately, you instruct Excel *not* to change cell references during the replication process. For example, in the calculations shown below, the overall income shown in cells C7, D7 and E7 are incorrect as the formulas copied from A7, that originally referred to the contents of cell A12, now refer to cells B12, C12 and D12 that contain no data.

To fix a cell address in a formula, you must add dollar signs. This is because Excel does not use relative cell references if you add a dollar sign $ in front of the column header letter and/or row header number of the particular cell.

Incorrect cell reference

Fig. 2.38 Absolute cell reference not set

set absolute cell references

1 Enter the formula as normal into the first cell.

2 For any cell whose position you want to fix, manually enter dollar signs in front of the column header letter and/or row header number.

 or

3 Click between **column letter** and **row number** of the cell address in the formula bar and press the function key **F4**. This circles through absolute, mixed and relative cell references.

4 When it shows the correct entry, click on the tick in the formula bar or press **Enter** to confirm the formula.

5 Replicate the formula in the normal way. All formulas will now refer to the same cell.

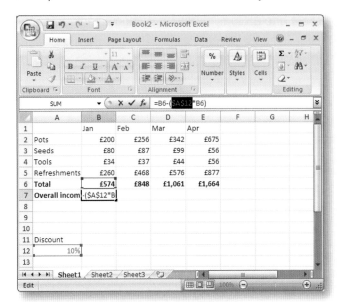

Fig. 2.39 Set absolute cell references

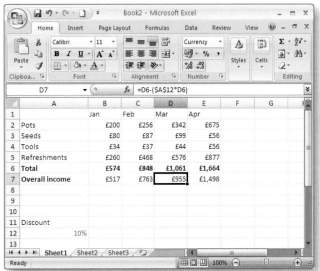

Fig. 2.40 Absolute cell reference replicated

Check your understanding 8

1 Open the file *Carpets* provided on the CD-ROM accompanying this book.

2 Use a formula to work out the area of Bedroom 1 (length x width) in sq. ft, and copy this down the column to calculate all the other areas.

3 One sq. ft. is equivalent to 0.0929 sq. metres. Type this figure in cell A9.

4 Using the cell address and *not* actual figures, work out the area of Bedroom 1 in sq. metres.

5 Copy this formula down the column to work out all the other areas.

6 Bedroom carpet costs £25.75 per sq. metre and the study carpet costs £15.25 per sq. metre.

7 Use a formula to work out the cost for carpeting Bedroom 1.

8 Now work out the cost for carpets in all the other rooms.

9 Format all prices to currency with 2 decimal places.

10 Format all other figures to 2 decimal places.

11 In row 6 add the heading *Total* and work out the total cost for all carpets.

12 Save the file as *Carpets calculated*.

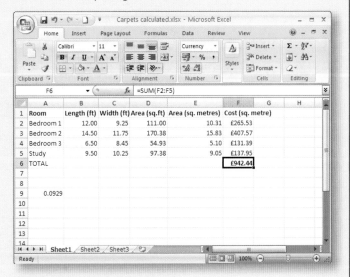

Fig. 2.41 Carpets calculated

Naming cells or cell ranges

A different way to refer to cells on a worksheet is to do so by setting and using a name that can be remembered more easily. Once named, you can type the name into a formula and it will be recognised in exactly the same way as a normal cell address.

For example, you could type

   =A1*B12

or rename B12 as *Costs* so that it becomes

   =A1 *Costs

Note that this is another way to use absolute cell addresses in a formula as only one cell (or cell range) can have the designated name. If you replicate a formula that contains a named cell, this name and the absolute cell address will be copied at the same time.

For example:

   =C5 *Costs (the contents of B12)

copied from Row 5 to Row 6 will become

   =C6*Costs (the contents of B12)

name a cell or cell range

1 Select the cell or range of cells.

2 On the **Formulas** tab, click on **Define Name**.

3 In the box that opens, type a name that begins with a letter and has no spaces. You will notice that, if the cell contains text or there is a text label nearby, that name will be offered by default.

4 The cell(s) being named will be displayed at the bottom of the box. If necessary, click on the red button to return to the worksheet to select a different cell/range.

5 Click on **OK**.

6 When you select the cell(s) on the worksheet, the name will show in the **Name** box.

7 When working with named cells, you will find that as you start entering the name into a formula, a list of possible names is presented to you. Double click on the correct name to add it to the formula automatically.

Name box

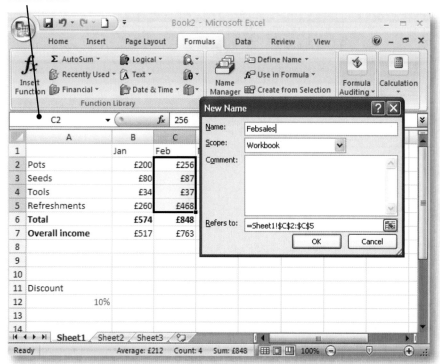

Fig. 2.42 Name cells

Check your understanding 9

1   Create a new spreadsheet.

2   Save as *Names*.

3   Enter the following data:

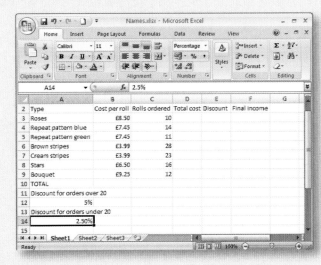

Fig. 2.43 Names

4   Widen all columns to display data in full.

5   Name cell A12 **Discount20**.

6   Name cell A14 **Discountbelow**.

7   Name the cell range C3–C9 **Orders**.

8   In D3 enter a formula to calculate the total cost for Roses paper (Cost per roll x Rolls ordered).

9   Replicate this formula to work out the total costs for all the other papers.

10   Using the correctly named cell, work out the discount for Roses paper (Total cost x Discount).

11   Enter formulas to work out all the other discounts.

12   Now use a formula to work out the Final income for Roses (Total cost – Discount).

13   Replicate this formula to calculate final income.

14   Finally, using the named cell range, enter a Sum function in C10 to work out the total orders received (=SUM(cell range)).

15   Format all financial data to currency with 2 decimal places.

16   Format the row heading and total for Rolls ordered in bold.

17   Save and close the file.

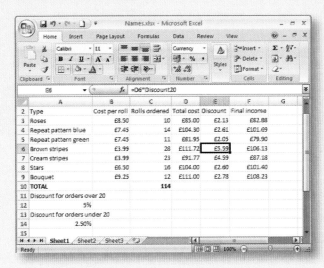

Fig. 2.44 Names completed

Functions

You know from Level 1 that you can use the **SUM** function to total figures in a range of cells. This is either typed in directly, where it takes the form **=SUM(cell range)** or you can enter it automatically by clicking on the Σ **AutoSum** button.

In the same way, common functions available from the drop-down arrow next to the **AutoSum** button include:

> **Average** – written as =AVERAGE(cell range) which works out the total divided by the number of entries
>
> **Maximum** – written as =MAX(cell range) which displays the highest value in a cell range
>
> **Minimum** – written as =MIN(cell range) which displays the lowest value in a cell range
>
> **Count** = written =COUNT(cell range) which totals the number of cells containing numeric data.

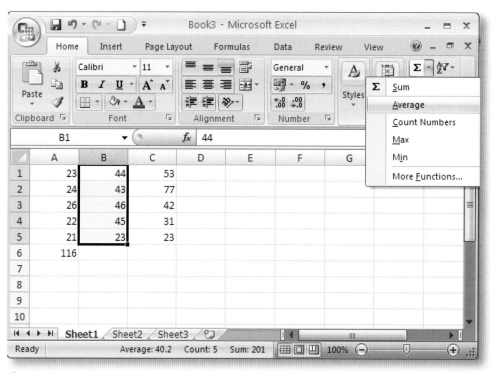

Fig. 2.45 AutoSum

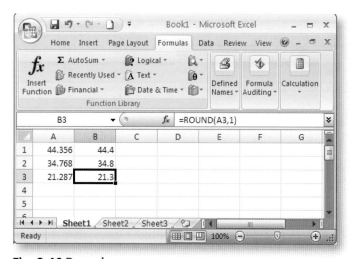

Fig. 2.46 Round

For Level 2, you need to be familiar with several other functions that can all be replicated down columns and across rows. They include the following:

ROUND

This rounds a figure to a set number of decimal places:

=ROUND (the actual figure/cell address, number of decimal digits)

Date

If you want to use the computer's built-in dates:

=TODAY () will display today's date

=NOW () will display both today's date as well as the time

IF

This returns one value if a condition is met, and another if it is not. To use this function, you must decide:

- what condition is to be met – for example, *a cell must contain a figure that is more than 20*, or *text = 'blue'*. (Text must be typed inside quote marks.)
- what to display if the condition is met – for example, it could show a figure, (e.g. *1*), or text (e.g. *'PASS'*).
- what to display if the condition is not met – this can be another figure or text entry. To display nothing, just enter double quote marks *" "*.

The function is written:

=IF(condition, entry if true, entry if false)

For example:

=IF(A1>10,'Pass','Fail')

or

=IF(A1='blue',1,0)

Note that you can use **IF** functions to help users input data into a spreadsheet. For example, you could add an **IF** function to instruct users to enter a name, date and so on into a neighbouring cell.

Acceptable operators for comparison include:

< less than
> more than
>= more than or equal to
<= less than or equal to
<> not equal to
= equal to

Nested IF function

Although you can use several **IF** functions repeatedly, you can also combine them into a single nested function. For example, you may want to distinguish between a range of different values so that you display the entry 'low' if a cell contains values between 0 and 10, 'mid' if it contains values between 11 and 20 and 'high' if it contains values over 20. This is written as follows:

=IF(A1<10,'low',IF(A1<20,'mid', 'high'))

Note that where there are several functions, they must all have closing brackets.

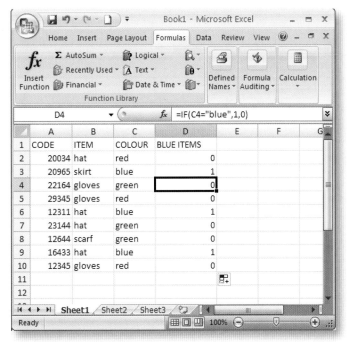

Fig. 2.47 IF function

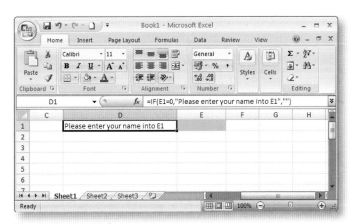

Fig. 2.48 Use IF for prompt

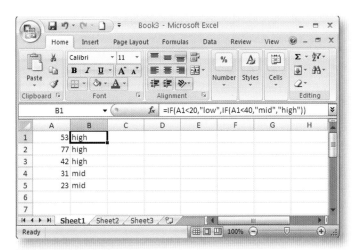

Fig. 2.49 Nested

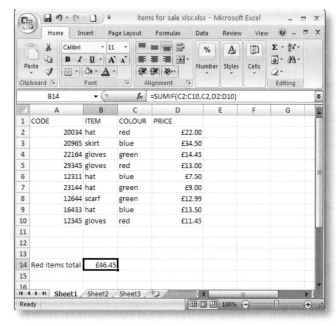

SUMIF – this will add together the values of cells that meet certain criteria.

To use this function, you must specify:

- the spreadsheet cells that will be compared
- an example of the criteria used for comparison
- the range of cells to total if the condition is met.

In the example below:

- the items to be compared are in cells C2:C10
- an example of the criterion, red, is in cell C2
- the entries to be totalled are in cells D2:D10.

Therefore the function is written:

=SUMIF(C2:C10,C2,D2:D10)

Fig. 2.50 SUMIF

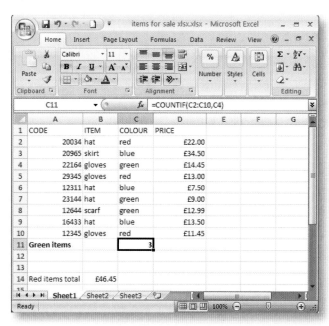

COUNTA

This counts cells containing any form of data, text or numerals. The function is written as:

=COUNTA(cell range)

and it will not count cells that are blank.

COUNTIF

This counts cells that meet certain criteria. The function is written as:

=COUNTIF(cell range, an example of the criterion)

In the example below, if you want to know how many items are green, you would enter the function as:

=COUNTIF(C2:C10,C4)

Fig. 2.51 COUNTIF

Check your understanding 10

1 Open the file *Functions* provided on the CD-ROM accompanying this book.

2 Complete the entry in cell D13 using the SUMIF function.

3 Complete the entry in cell C15 using the COUNTIF function.

4 In column F, use an **IF** function to display 'Yes' if Unit prices are equal to or less than £14.99 and 'No' if they are above this.

5 Use a function to display today's date in cell A17.

6 Save as *Functions completed*.

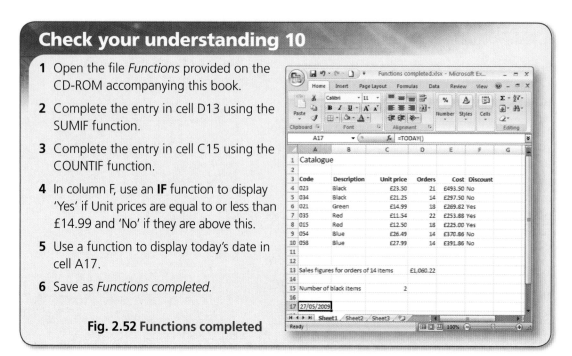

Fig. 2.52 Functions completed

The function library

If you are ever unsure about correctly structuring a function, you can use the function library. Either click on an appropriate command for a list of relevant functions or click on the **fx** command in the **formula** bar or on the **Formulas** tab to open the dialog box.

Search for the function you want to use and enter the correct cell ranges or other entries suggested to complete the function correctly.

Worksheet

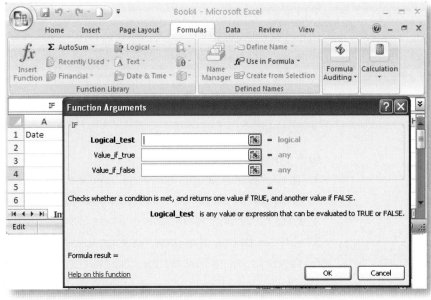

Fig. 2.53 Function box

names

When several sheets in a single workbook are in use and you need to move between them to enter or edit data, it is helpful to use relevant names, rather than leave the default as Sheet1, Sheet2 and so on.

rename a worksheet

1 Double click on the **sheet** tab.

or

2 Right click and select **Rename**.

3 Enter a new name over the name that will now be highlighted.

Check your understanding 11

1 Reopen *Names*.

2 Change the name of Sheet1 to *Paper*.

3 Save and close the file.

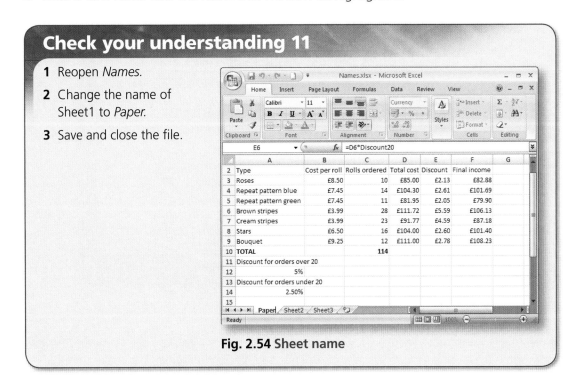

Fig. 2.54 Sheet name

Copying and linking

If you want to copy text entries, values that won't change or even graphics from one spreadsheet to another, it is simple to use **Copy and Paste**. However, it is more complex where the data changes or involves formulas.

As well as performing calculations within a single worksheet, you can incorporate data from different worksheets or even other workbooks. When you do so, you must include the sheet name and, if relevant, workbook name in the formula so that Excel can perform the calculation accurately.

link spreadsheets

1 To display a value that is in a cell on another sheet, type = and then navigate to the sheet and click on the cell containing the entry you want to copy.

2 Press **Enter** and its contents will be copied across.

3 For calculations, start creating a formula in the normal way.

4 When you need to enter the cell reference for cells on a separate sheet, click on or select the cells with the mouse as you build up the formula.

5 You will see that the cell addresses from a different worksheet are prefixed by the sheet name (either the default or a new name it has been given) and an exclamation mark.

6 Cells from a different workbook are prefixed by the filename in square brackets and then the sheet name, an exclamation mark and the absolute cell references.

7 You can type in the cell addresses rather than select them with the mouse, as long as the sheet or workbook details are included and entered in the accepted format.

Separate workbook name

Sheet in the same workbook containing the original data

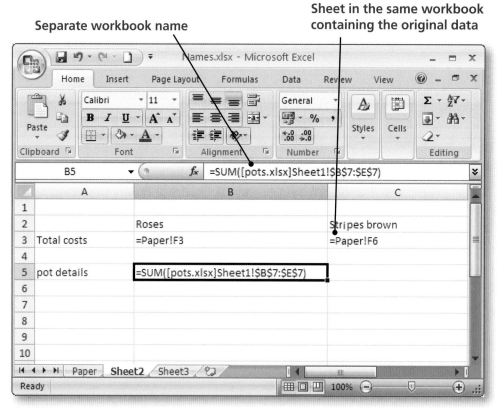

Fig. 2.55 Linked worksheets

If you simply want to copy across the contents of a particular cell, you can use **Copy and Paste**. However, if the original entry changes, this will not be reflected in the copy. To make sure that both are updated, you can link the two sheets using a Paste special option.

You should also use the linking option if you want to copy formulas. Otherwise they will refer to the wrong cells.

link worksheets using Paste Link

1 Select the cell(s) you want to copy on one worksheet.

2 Click on **Copy**.

3 Click on the first receiving cell.

4 Click on the drop-down arrow below the **Paste** command on the **Home** tab.

5 Select **Paste Link**.

6 The cell contents will appear but, if you look in the formula bar, you will see the name of the original sheet and cell address rather than the value or text that has been copied across.

7 Pasted formulas will still refer to the appropriate cells on their original sheet.

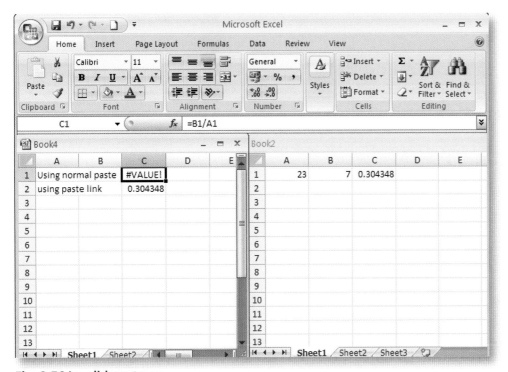

Fig. 2.56 Invalid paste

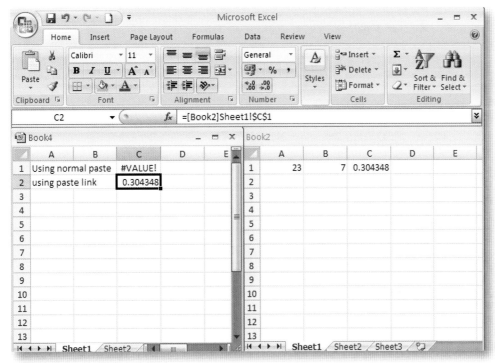

Fig. 2.57 Paste link

8 If in future you change the entry on the original cell, this will be reflected in the copy.

Check your understanding 12

1 Open the file *Villas* provided on the CD-ROM accompanying this book.

2 Start a new workbook named *Countries*.

3 Enter the following data:

Fig. 2.69 Countries

4 Enter the correct formula in A3 to work out the total visitors to France using data from the file *Villas*.

5 Work out totals for all the other countries.

6 Note down the total visits to Sweden in cell A6.

7 Close *Countries*.

8 Now change the entry in the *Villas* file for numbers visiting Sweden in May to 25.

9 Reopen *Countries* and check that the total for Sweden has changed.

10 Save and close both files.

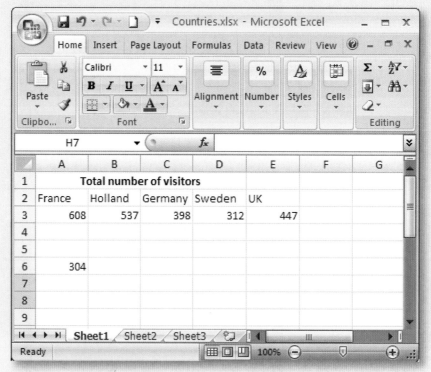

Fig. 2.58 Countries final

Sorting

Having entered all the records in your database, you may want to reorder the details to help you locate particular records for easy reference. For example, you may want to display the records in alphabetical order or from the highest to lowest price, size or weight.

Excel enables you to sort records using a range of different criteria but when using the application in this way, it is important to remember that it is possible to select and sort data within single columns. As this will split up the information related to an individual record, you need to take care to select *all* the data before carrying out a sort. (In Excel 2007 you should see a warning message if you make this mistake, but do not rely on this happening.)

sort records

1 Select all the data, including column headings.

2 Click on the A–Z button on the **Data** tab to sort alphabetically or in ascending order (from lowest/earliest to highest/latest). The sort will be based on data in the first column.

3 Click on the Z–A button to sort in descending, reverse order, again based on the first column.

4 To select a different column on which to base the sort, or to set different levels of sort, click on the **Sort** command.

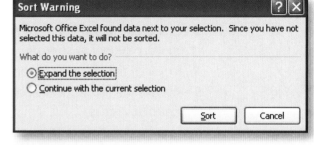

Fig. 2.59 Sort warning

5 In the dialog box, select the first column on which to base the sort and then select the order – for example, A–Z or lowest to highest.

6 Note that there is a tick in the checkbox showing headers have been selected. (If you remove the tick, you would have to sort by column letters rather than categories and the headings would be included in the sort.)

7 To set a second order sort, click on the **Add Level** button and repeat the process.

8 When you have completed all the levels, click on **OK** to reorder your data.

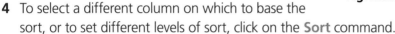

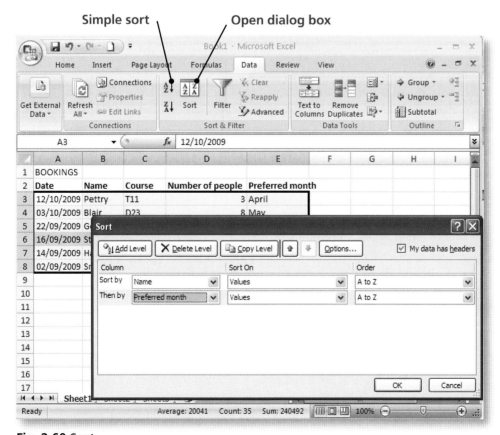

Fig. 2.60 Sort

Searching for records

The main value of a database is to be able to locate records matching certain criteria. For example, in the above database you could search for any names beginning with *H*, all bookings made in September, or number of people under 5.

When searching you are actually filtering the records using the **AutoFilter** function. This hides all records that do *not* match your chosen criterion, leaving only matching records visible. You can continue to filter this subset of the database or display all the records again.

search the records

1 Select all the records including column headings.

2 Click on the **Filter** button. This adds a small drop-down arrow to each category.

3 Click on the arrow in any category box. At this stage you could sort all the records based on this category by selecting a sort order.

4 To search the records, click on the arrow in the appropriate category.

5 Remove the tick in the **Select All** box and then click on a single checkbox to display all matching entries.

6 To set your own criteria, rest your mouse on **Text** or **Number Filters** and choose an option such as **Equals**, **Greater than** and so on, or click on **Custom Filter** if none of these are appropriate.

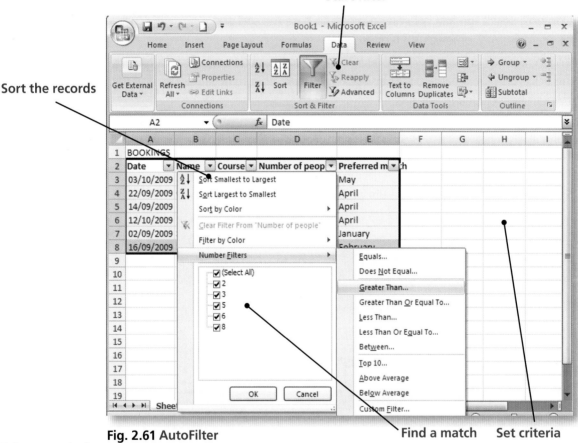

Sort the records

Start filter

Fig. 2.61 AutoFilter

Find a match   **Set criteria**

Select or type in

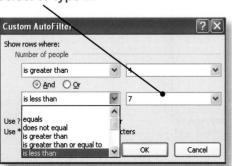

7 In the dialog box, select the type of filter you want to apply and either select or type in your criteria in the boxes provided.

8 Click on **OK** and only matching records will be displayed.

9 Filter this set of records or click on the **Clear** or **Filter** button to display all the records.

Fig. 2.62 Custom filter

Check your understanding 13

1 Open the file *Rental properties* provided on the CD-ROM accompanying this book.

2 Sort the records in descending order of Rental income.

3 Now sort the records in ascending order of Bedrooms.

4 Use the AutoFilter to display only those properties that are flats.

5 Remove the filter and search for all properties with three or more bedrooms.

6 Save and close the file.

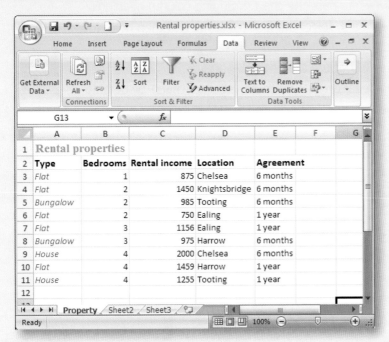

Fig. 2.63 Sort bedrooms

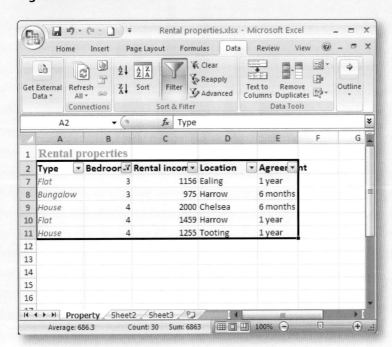

Fig. 2.64 Filter bedrooms

Graphs and charts

To create a basic chart, you can either use a shortcut or follow the steps offered by the chart wizard.

When the chart appears, you will find three different **Chart Tools** tabs available:

- **Design** – enables you to change chart type and move the chart
- **Layout** – offers a range of titles, labels and other display options
- **Format** – lets you format selected elements of the chart.

create a chart using the shortcut

1 Select the data on which to base the chart, including column headings.
2 To select non-adjacent data, select the first cell range and then hold down **Ctrl** as you select further cells.
3 Press the function key **F11** at the top of the keyboard.
4 A column chart will appear on its own sheet, labelled Chart1 by default.
5 Your data will still be available – usually on Sheet1.

Fig. 2.65 Shortcut chart

create a chart using the wizard

1 Select the data including column headings.
2 On the **Insert** tab, click on the basic type of chart you want to create – for example, pie, column, bar or line.
3 Now select the chart style you prefer, e.g. exploding pie chart or line chart with markers.
4 A chart will appear on the same sheet as the data.

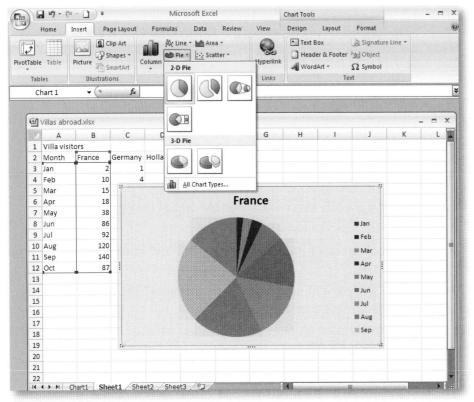

Fig. 2.66 Insert chart

Axes and titles

When a chart first appears, it will show the data series as coloured bars or pie slices and a legend identifying each series. There may also be a chart title based on one of the column headings.

To display the data clearly, you may need to add or edit the main and axes titles.

add a title

1   Click on the **Chart Tools – Layout** tab.
2   Click on the **Chart Title** command and select your preferred position for the title such as on (overlay) or above the chart area.
3   A default entry will appear in a box that is already selected.

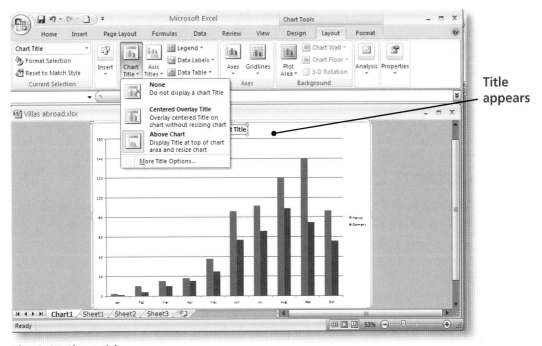

Title appears

Fig. 2.67 Chart title

add Axis titles

1 Click on the **Axis Title** command and select primary **Horizontal (X)** or **Vertical (Y)** axis.

2 Again, select its position and style.

Note that sometimes the chart will show the data displayed against the wrong axis. If this happens, click on the button on the **Chart Tools – Design** tab to switch between columns and rows.

edit a title

1 Select an unwanted title and press **Delete** or click on the relevant **Axis** or **Chart Title** command and select **None**.

2 To change an entry:

   a Click in the title box and use the keyboard to edit the text in the normal way.
     or
   b Select the title box and then enter your choice of text into the **Formula** bar. Click on the tick to replace the title.

3 To format the title, select the box and use the normal formatting tools on the **Home** tab to change font, font size, emphasis and so on.

4 To move the title to a different position, select the box and then drag it when the pointer shows a four-way arrow.

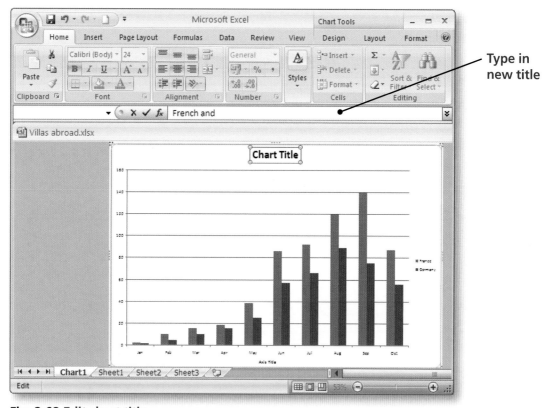

Fig. 2.68 Edit chart title

Changing chart type

When using the shortcut, or if you change your mind about the display, you can change the chart type once a chart has appeared. Commonly:

- a **histogram** (columns that touch) shows discrete frequency distribution
- a (vertical) **column** or (horizontal) **bar chart** shows observations over time or under different conditions
- a **pie chart** compares relative amounts at a point in time, with percentage values seen as a slice of a pie
- a **line chart** shows trends over time
- a **scatter plot** shows the distribution of data points along one or two dimensions.

change chart type

1 Select the chart.

2 On the **Chart Tools – Design** tab click on **Change Chart Type**, or right click for this option.

3 Select a new type and style of chart from the gallery that will appear.

4 Click on **OK** to return to the chart.

Change chart

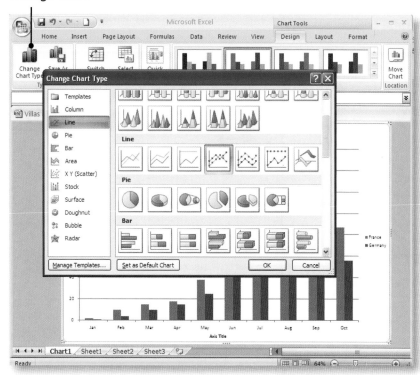

Fig. 2.69 Change chart type

Legends and labels

For some charts it is helpful to add labels to data series such as pie slices. You may also prefer to remove the legend once the data is clearly identified.

edit the legend

1 Click on the legend box and press the **Delete** key to remove it.

2 Drag it to a new position using the mouse.

3 All these options are also available from the **Legend** command on the **Chart Tools – Layout** tab.

4 To format the legend, select the box and use tools on the **Home** tab.

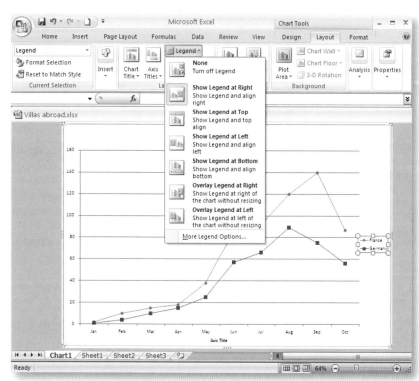

Fig. 2.70 Chart legend

add data labels

1 Select the chart.
2 Click on the **Data Labels** command on the **Chart Tools – Layout** tab and select a position for the labels or click on **None** to remove them.
3 To choose what labels to display, click on **More Data Label Options**.
4 You can now display the values, headings (categories) and/or data as percentages.
5 Add leader lines if you want to join labels to their data.
6 Back in the chart, drag any label to a new position if it is not clearly displayed, and format using tools on the **Home** tab.

Add labels

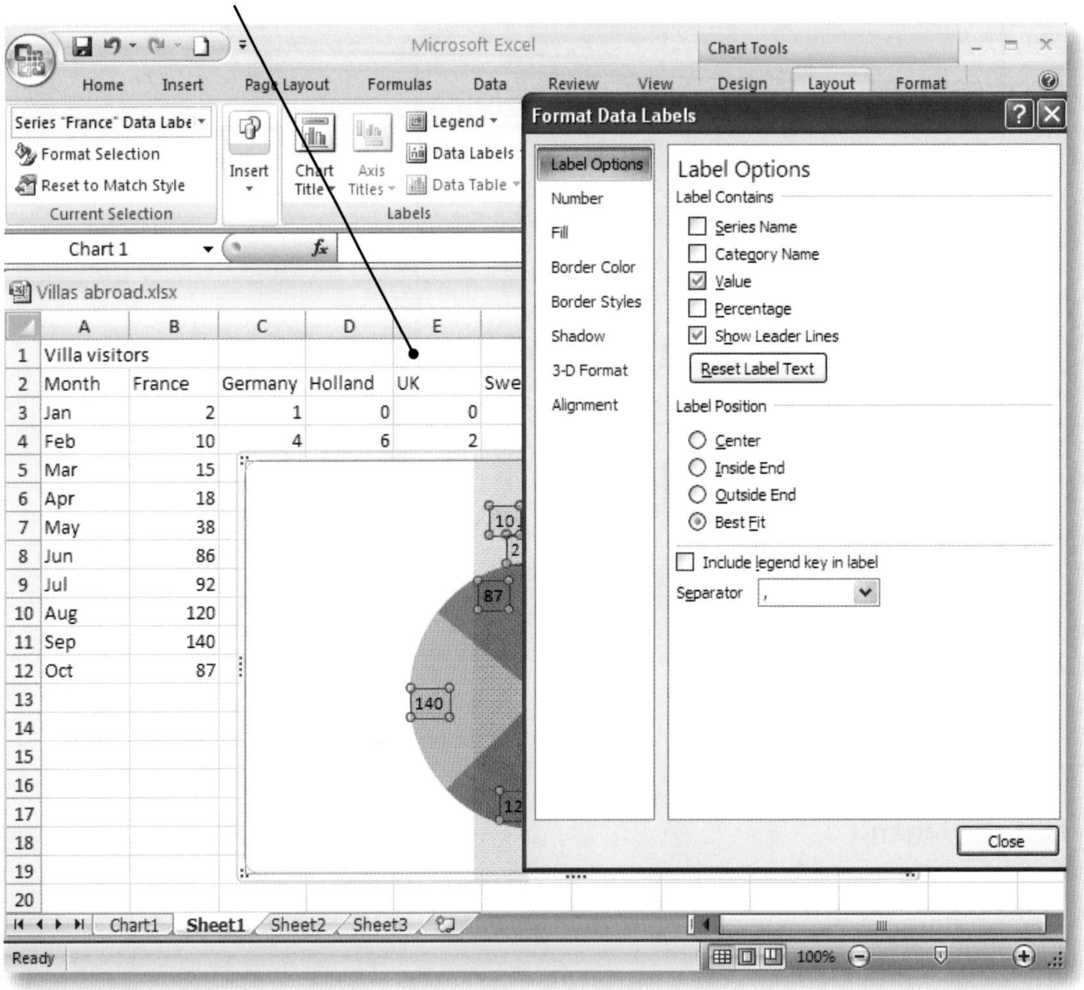

Fig. 2.71 **Data labels**

Rename sheet

Fig. 2.72 **Move chart**

Chart position

If your chart appears on the same sheet as the data, you can move it to its own sheet. This can be named to help you locate the chart in future.

move a chart

1 Select the chart.
2 Click on **Move Chart** on the **Chart Tools – Design** tab.
3 Type in a name for the sheet.
4 Click on **OK**.

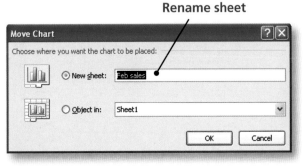

Check your understanding 14

1 Open the file *Sales figures* provided on the CD-ROM accompanying this book.

2 Create a 2D pie chart showing February figures for the three locations and place the chart on the same sheet as the data.

3 Enter the chart title *February Sales*.

4 Add data labels to show percentages and locations.

5 Remove the legend.

6 Position the chart title in the top right-hand corner and change the font to size 16.

7 Reposition any data labels to make them clearer.

8 Now change to a 2D bar chart.

9 Move this to its own sheet named *Feb sales*.

10 Remove the data labels.

11 Redisplay the legend.

12 Add the Axis titles *Sales* and *Locations*.

13 Reposition the chart title centrally.

14 Save as *Sales charts* and close the file.

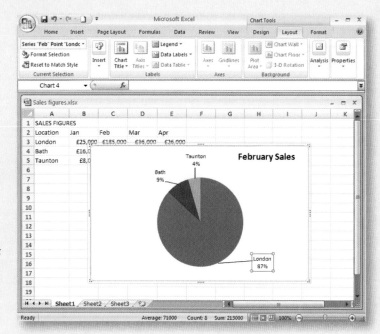

Fig. 2.73 Feb sales step 1–7

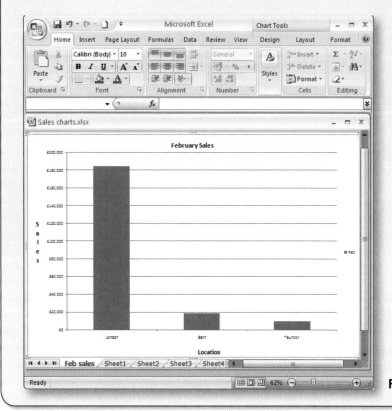

Fig. 2.74 Sales chart

113

Changing upper and lower limits

Where the maximum and minimum values are not appropriate, you can change the default settings and set a different scale for the chart.

change upper and lower limits

1 Select the relevant axis by clicking on a value.

2 Click on the **Chart Tools – Format** tab.

3 Make sure *Vertical axis* is showing in the **Chart Elements** box. If not, select it from the drop-down list.

4 Click on **Format selection**.

5 In the dialog box, click on **Fixed** to set a new maximum and/or minimum value and enter the new figure in the box.

6 You may need to change the major units if the spacing/interval between values is wrong after changing the scale.

7 Click on **OK** to return to the chart.

Chart element

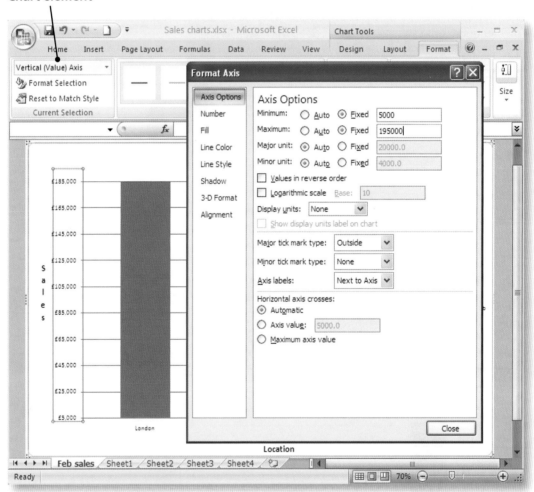

Fig. 2.75 Scale

Emphasising chart elements

As well as adding colours to the chart or plot area, you can use colour for different lines or columns/bars to help identify each data series more clearly when you are comparing data, or if the default colours are not appropriate. There are also other changes you can make such as increasing line thickness in a line graph, adding line markers or exploding a pie chart slice.

change fill colours and lines

1   Select the element such as plot area, line or column.
2   Right click and select the **Format** option, or click on the **Format Selection** command on the ribbon.
3   When the dialog box opens, select the type of formatting to apply and then choose from drop-down colour palettes or line styles that are offered.
4   Click on **OK** to apply the new format.

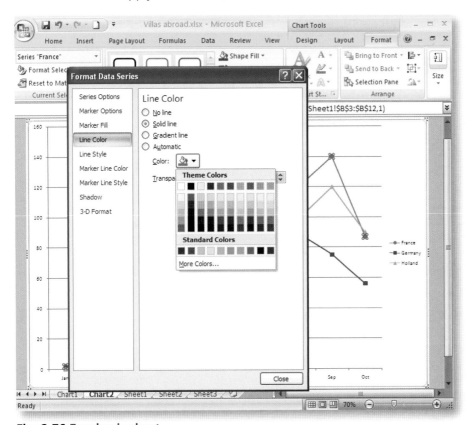

Fig. 2.76 Emphasis chart

explode an individual pie slice

1   Click on the pie slice to select it.
2   You may need to click again to make sure it is selected separately from other slices.
3   Drag the slice outwards with the mouse.

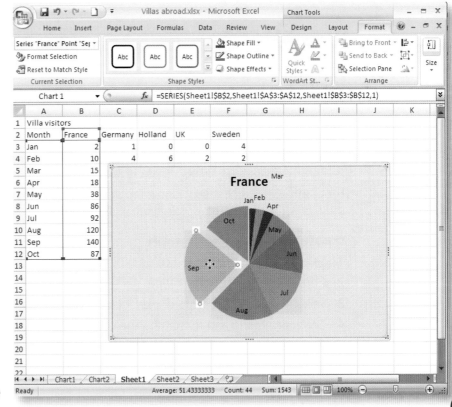

Fig. 2.77 Explode

Check your understanding 15

1 Open the file *Weight loss* provided on the CD-ROM accompanying this book.

2 Create a line chart to display the weight over time for Sally and Delia.

3 Add a chart title *Weight Loss*.

4 Add axis titles *Pounds in weight* and *Time* and change these to font size 10.

5 Shade the plot area pale yellow.

6 Increase the line width for Sally's data.

7 Change the scale from 135 lb to 165 lb.

8 Add data labels to show the weight (values).

9 Position the chart on its own sheet labelled *Weight Loss*.

10 Save and close the file.

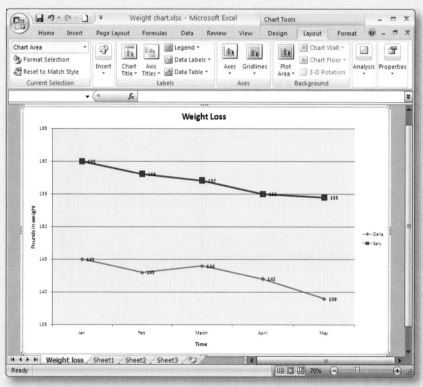

Fig. 2.78 Weight Loss

XY (Scatter) graphs

One type of graph you may want to create is an **XY (scatter) graph**. This is used to show irregular data more clearly than a line graph and compares variables on both the X and Y axis. For example, it can be used to show measurements of growth taken over time. You can either show only the data points or join them with a line.

create an XY (scatter) graph

1 Select the data but not the column headings.

2 Choose the **Scatter chart** type.

3 Select the style of chart to display – i.e. with or without connecting lines.

4 Continue editing and formatting the chart as normal. You will need to add titles and possibly label the data series to make sure the display is clear.

Check your understanding 16

1 Open the file *Hot days* provided on the CD-ROM accompanying this book.

2 Create a scatter graph.

3 Give the chart the title *Days temperature over 80 degrees*.

4 Label the Y axis *Number of days*.

5 Label the X axis *Year*.

6 Save as *temperature*.

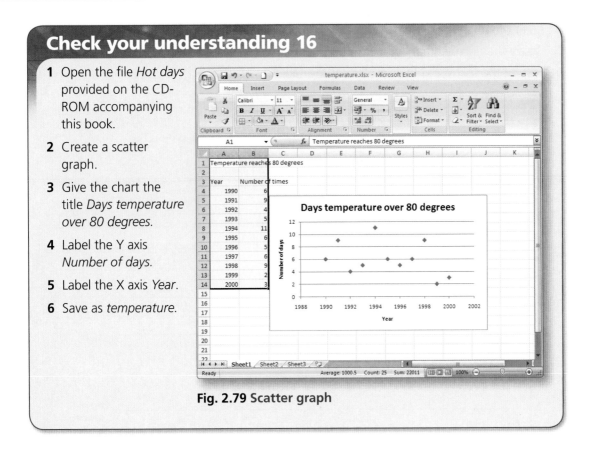

Fig. 2.79 Scatter graph

Text boxes

One way to label data is to add a text box and position it over a line or column. To do this, you must use the shapes option on the **Insert** tab, which was explained in more detail in Unit 001.

add a text box

1 Click on the **Insert** tab.

2 Click on **Shapes**.

3 Click on the **Text Box** option and then draw a text box onto the chart.

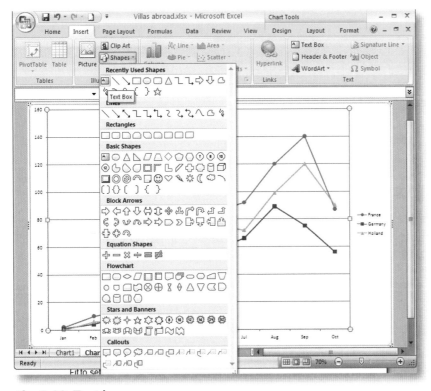

Fig. 2.80 Text box

4 Enter the text where the cursor is flashing.

5 Resize the box by dragging a boundary outwards.

6 Format the text using tools on the **Home** tab.

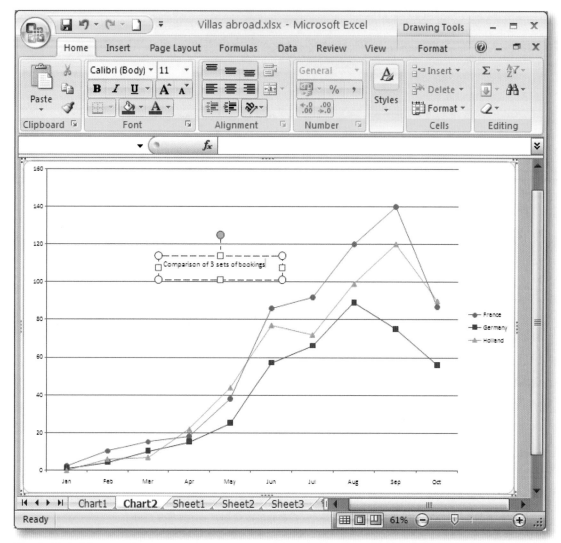

Fig. 2.81 Add a text box

Check your understanding 17

1 Open any file containing a chart.

2 Add a label to the chart using a text box.

Multiple charts

As well as setting different chart types, it is possible to combine different types of chart. This can be useful if you are comparing a number of different sets of data – for example, performances against a benchmark – and want one series to be a line and the rest in columns or bars.

create a multiple chart

1 Select all the data.

2 Create the basic chart, e.g. a column chart.

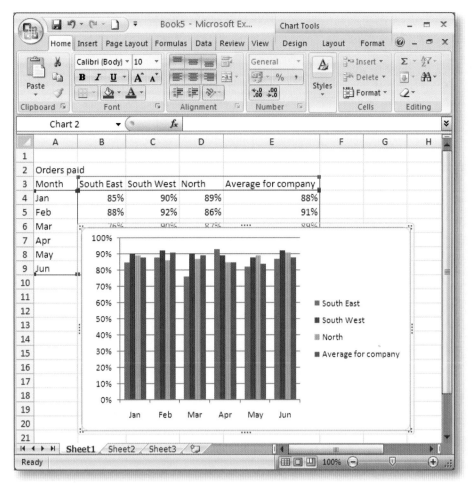

Fig. 2.82 Multiple step 1

3   When it appears, right click on the data series you want to show as a line.
4   Select **Change Series Chart Type**.

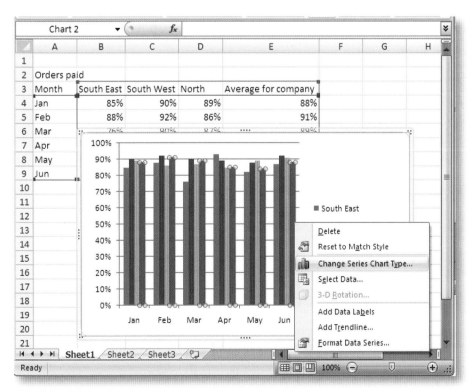

Fig. 2.83 Multiple step 2

5   Select the chart type, e.g. a line from the gallery and click on **OK**.
6   This data series will now appear as a line against the rest of the data still in columns.

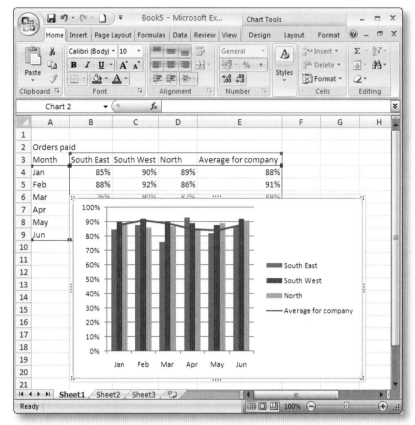

Fig. 2.84
Multiple step 3

Check your understanding 18

1   Open the file *Paid* provided on the CD-ROM accompanying this book.
2   Create a multiple chart on its own sheet where the *Average for company* data is a line
    and the rest of the data is displayed as a bar chart.
3   Increase the line weight to make it stand out.
4   Save and close the file.

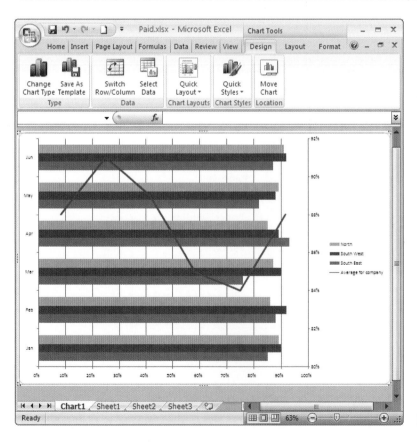

Fig. 2.85 Paid

Printing

You will already know how to print spreadsheet data in table format, and how to increase the number of copies to print out. To change various other settings, you need to go to the **Page Layout** tab, the **Page Setup** dialog box reached from this group or print options via the **Office** button.

Some of the choices you may need to make include:

- setting the number of pages the spreadsheet will spread across
- printing formulas rather than values
- printing a section of the worksheet
- printing in colour or black and white
- printing column or row headings on every page
- printing with or without gridlines visible
- printing charts.

Before printing, always check in **Print Preview** to make sure you print exactly what you want to see. You may need to change from Portrait to Landscape or change the paper size.

preview your worksheet

1 Click on the **Office** button.
2 Hover over **Print** and select **Print Preview**.
3 Either go directly to the **Print** or **Page Setup** dialog boxes or close the preview and return to the spreadsheet.

Groups on the **Layout** tab link to different sections of the **Page Setup** dialog box.

select which parts of a spreadsheet to print

1 To print a limited cell range, select the cells.
2 Click on the **Office Button** and select **Print** to open the **Print dialog box**.
3 Click on the radio button next to **Selection** in the **Print what:** box.
   or
4 Click on **Print Area – Set Print Area** before printing.
5 Note that the data may print out onto more than one page.
6 Take off this setting by selecting **Print Area – Clear Print Area**.

Fig. 2.86 Set print area   **Open Page Setup**

print column or row text entries on multiple pages

1 Click on **Print Titles**.
2 Click on the **red** button in the **Print Area** window to move to the spreadsheet.
3 Use your mouse to select the headings you want to repeat and then click on the **red** button again.
4 The absolute cell references of the cells you selected will now be displayed in the box.

Print Titles

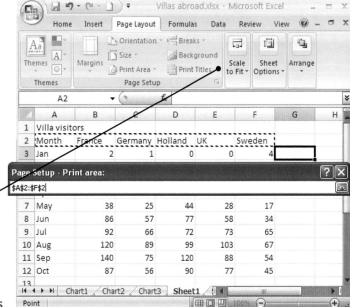

Fig. 2.87 Print column entries

print gridlines and/or headings

1 Click to place a tick in the **Print** checkbox under **Headings** or **Gridlines** in the **Sheet Options** group if you want to print these out. Click again to remove the tick if you don't want them to print.

2 With headings ticked, you will print out column letters and row numbers as well as your data.

   or

3 Click on the **Sheet** tab in the **Page Setup** dialog box to locate the relevant checkbox.

Click to display on printout

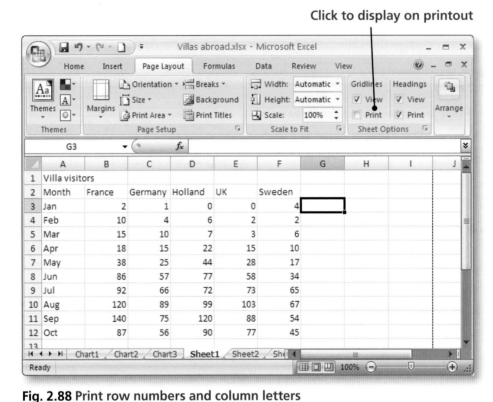

Fig. 2.88 Print row numbers and column letters

set print size

1 Open the **Page Setup** dialog box.

2 On the **Page** tab, click on the **Fit to:** radio button and leave both measures at 1 or increase the number of acceptable pages.

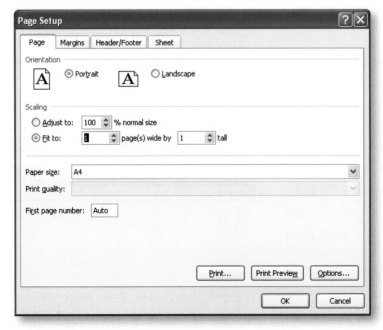

Fig. 2.89 Fit to one page

Printing formulas

Where you want to print out the underlying formulas used for any calculations, you need to change to this view before printing.

1 Click on the **Formulas** tab.

2 Click on **Show Formulas**.

or

3 Hold **Ctrl** and press the ¬ key (to the left of **1** on the spacebar).

4 Repeat either of these to return to showing values.

5 When formulas are visible, the spreadsheet will widen. You will see a dotted line marking the edge of the page.

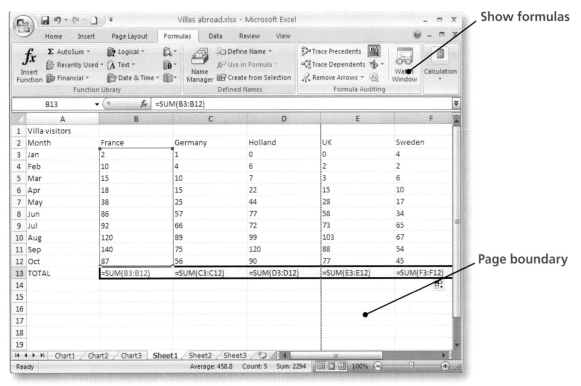

Show formulas

Page boundary

Fig. 2.90 Show formulas

Printing charts

- For charts on the same sheet as the data, click on the chart to print it alone.
- Leave a chart unselected to print chart plus data.
- For charts on their own sheet, simply open the sheet and this will be treated as the active sheet.

Charts are automatically coloured on screen in Excel 2007 but the data series may not be clearly identified when they are printed in black and white. To make sure you can compare data when printing without colour, use the facilities in Excel to add patterns.

1 Select the chart and open the **Page Setup** dialog box.

2 Click on the **Chart** tab.

3 Select **Black & White** and click on the **Preview** button.

4 Click on **OK** to print in this format.

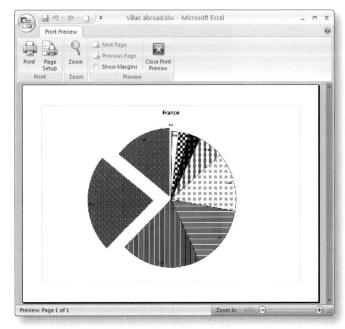

Fig. 2.91 Black & White chart

Check your understanding 19

1 Open *Weight loss*.

2 Print the chart on its own in black and white making sure the data series are clearly identified.

3 Now print only the data related to Sally's weight.

4 Finally, print out the data to include column and row headings.

5 Close the file.

Importing data files

Excel can open and save files in a number of different formats, and this includes the generic data file type **CSV** (**Comma Delimited** or **Comma Separated Value**). As the name indicates, data is separated by commas and this type of file can be opened by Word as well as Excel.

When opened in Excel, or when saving a spreadsheet in CSV format, the data will look like a normal spreadsheet but the formatting will be simplified and any underlying formulas will be lost. This means that, if figures are changed, calculations based on them will not be updated. CSV files can also only contain a single worksheet.

Having opened a CSV file, it is important to save it as an Excel file so that you can retain any formulas or formatting applied in future.

open a CSV file

1 Click on the **Office** button.

2 Click on **Open**.

3 Navigate to the folder containing the file.

4 In the dialog box, click on the **Files of type:** box and select **All Files**.

5 Select the target file.

6 Click on **Open**.

Fig. 2.92 Open CSV file

save a CSV file as an Excel file

1 Click on the **Office** button.

2 Click on **Save As**.

3 In the **Save As** dialog box, click on the drop-down arrow in the **Save as type:** box and select the first option – **Excel Workbook**.

4 Edit the file name and change the save location if necessary.

5 Click on **Save**.

6 Note that the CSV filename will now be the name for Sheet1.

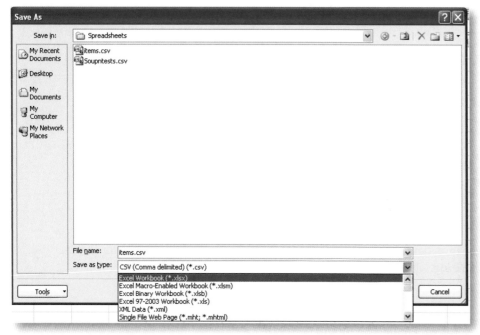

Fig. 2.93 Save CSV file

CLAiT Assignment

CLAiT Plus Spreadsheet Assignment

You will need the following files:

- **hiring** in .csv format
- **colours** in .csv format
- **employees** in .csv format
- **turnover** in .csv format

Task 1

1. Open the .csv file **Hiring** and save it in your software's normal file type using the file name **Clothes**.

2. Apply the following formatting:

   a. Merge and centre the title across columns A – I.

   b. Change the title font to a large size, e.g. font size 16.

   c. Format the font for all column headings to a medium font size, e.g. font size 12 except for the cell containing the heading **Hiring** which should be bold, font size 13.

3. Apply the following alignment to column headings: vertical – centre, horizontal – centre.

4. In the cell containing the text **Hiring statistics for London**, apply a text wrap so that the entry appears on 2 lines without any words being split. Format the font to bold, size 13.

5. In the **Weekly Rate** column, format the numeric data to currency to display a £ symbol and 2 decimal places.

6. Wrap the text for the column heading **Number of items** and make the column narrow, still displaying the heading on 2 lines.

7. Make the following amendments to the spreadsheet:

   a. Below all the data, enter a new heading in column A containing the label TOTAL ITEMS IN THE LONDON SHOP. Wrap the text so the entry is on 2 lines.

   b. Delete the row containing all the data for March as the shop was shut then.

   c. Insert a column between **Number of items** and **Weekly rate**.

   d. Give it the heading **Codes** and enter the following codes for each item:

Evening dress	ED
Evening suit	ES
Bow tie	BT
Summer ballgown	SB
Wedding dress	WD

   e. At the bottom of the **Number of items** column, enter a function to total the number of items held in the London shop.

   f. Copy this entry to below the label TOTAL ITEMS IN THE LONDON SHOP.

8. Now you need to calculate whether the shop will need to borrow costumes from another outlet. If more than 85% of items are hired out in any month, they will have to borrow.

   a. In the **TOTAL OUT** column, use a function to total all hires for February.

   b. Replicate this formula to total hires for the other months.

9. In the **Borrowing** column for February, use a function to display whether the shop needs to borrow or not. Use an absolute cell reference. If the figure for **Total Out** is less than 85% of the figure for **Total items** display **Borrow**, otherwise display nothing.

10 Replicate this formula for all the other months.

11 Make sure all data is fully displayed.

12 Now work out the income for February. You will need to create a long addition and make use of absolute cell references. Multiply totals for each type of item by their weekly rate and add them together.

13 Replicate the formula down the column to work out the income for all the other months.

14 Format the income results to currency with zero decimals.

15 Save the file with the same filename **Clothes** and close the file.

Task 2

1 Open the .csv data file **Colours** and save it in your software's normal file format with the name **Favourites**.

2 Create a line-column graph to plot the **multicoloured** hired clothes as a line against the figure for each colour as a column. Ensure data labels showing the colours are displayed on all except the line. Create the graph on its own worksheet.

3 Title the graph: **Favourite Colours**.

4 Label the X-axis: **Years**.

5 Label the Y-axis: **Numbers hired**.

6 Ensure the data is displayed against the appropriate axis.

7 Retain a legend.

8 Format the Y-axis as follows:

    a.   minimum value – 9

    b.   maximum value – 50

    c.   interval – 15.

9 Add your name and the filename as a header.

10 Save the file keeping the name **Favourites**.

11 Print one copy of the graph, ensuring the data is distinctive and will be clearly identified.

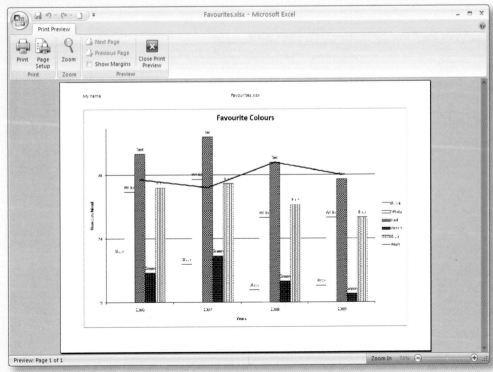

Spreadsheet 1

12 Close the file.

Task 3

1   Open the .csv data file **Employees** and save it in your software's normal file type using the name **Workers**.

2   Sort the data in descending order of **Rate per hour**.

3   Now make the following amendments:

    a.   Hilda's details have changed so the figures should now be as follows: Location **Aberdeen**, Contact **016645673452**, Hours available 28.

4   Filter the data to find all workers who can work 30 or more hours.

5   Below the last entry in the column labelled **Rate per hour**, use a function to count the number of workers whose rate is £9.25.

6   Name the cell containing the result of the count as **Mainrate**.

7   Add your name and the date as a header.

8   Ensure gridlines and row and column headings will be displayed when printed.

9   Save the file keeping the same filename.

10 Maintaining the filter, print a copy of the spreadsheet in landscape orientation showing the formulae and filtered data on one page.

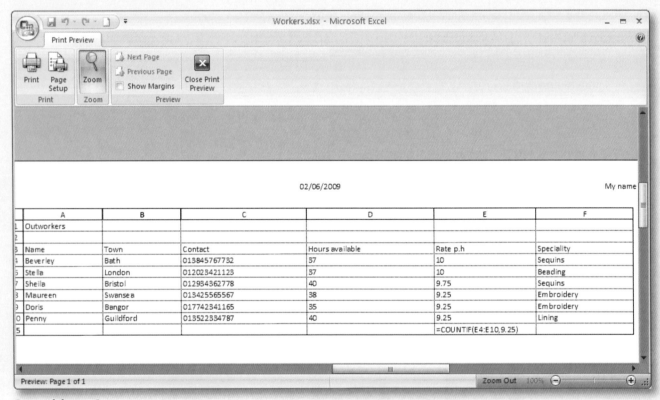

Spreadsheet 2

Task 4

1   Keeping **Workers** open, reopen the file **Clothes** saved earlier.

2   Copy the contents of the named cell **Mainrate** showing the results of the count into the cell just below where you copied total sales in the London shop.

3   Save the file **Clothes** to update any changes.

4   Print a copy of the spreadsheet in landscape orientation showing the figures.

Maisie's Clothes Hire Company

Hiring statistics for London

Type	Codes	Number of items	Weekly rate
Evening dress	ED	10	£150.00
Evening suit	ES	12	£210.00
Bow tie	BT	8	£25.00
Summer ballgown	SB	16	£80.00
Wedding dress	WD	9	£150.00
TOTAL ITEMS		55	

Hiring

Month	ED	ES	BT	SB	WD	TOTAL OUT	Borrowing	Income
Feb	6	8	3	0	4	21		£3,255
Apr	5	4	6	5	7	27		£3,190
May	10	10	8	11	9	48	Borrow	£6,030
Jun	10	11	8	16	9	54	Borrow	£6,640
Jul	9	10	7	14	9	49	Borrow	£6,095
Aug	7	6	4	12	8	37		£4,570
Sep	4	8	2	14	9	37		£4,800
Oct	6	5	5	10	6	32		£3,775
Nov	4	5	5	10	6	30		£3,475
Dec	7	4	6	8	4	29		£3,280

TOTAL ITEMS IN THE LONDON SHOP

55

3

Spreadsheet 3

5   Display the formulae.

6   Hide the four columns headed **ES – WD**.

7   Print a copy of the spreadsheet showing the formulae.

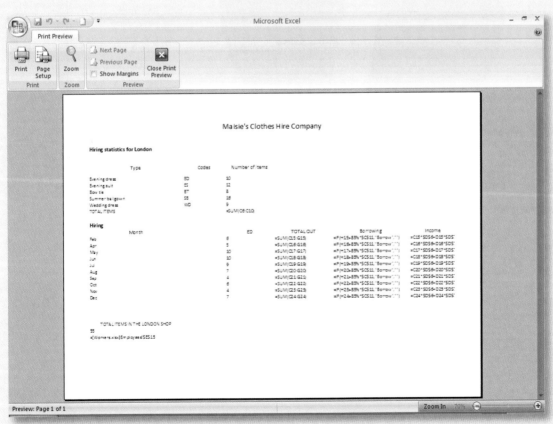

Spreadsheet 4

8   Save as **Finalclothes**.

Task 5

1 Open the .csv data file **Turnover** and save it in your software's normal file type using the name **Income**.

2 Create an exploded pie chart on its own sheet that will display the income for all the shops except **Clothes for hire**.

3 Title the chart **Annual Shop Income**.

4 Each sector must be clearly labelled with the figures as percentages.

5 Display a legend.

6 Pull out **Maisie's** sector of the chart.

7 Alongside this sector, add a text box containing the words **Highest performer**.

8 Ensure the text box does not touch or overlap the pie chart sectors or legend.

9 Add your name and the date as a footer.

10 Save the file keeping the same filename.

11 Print one copy of the chart.

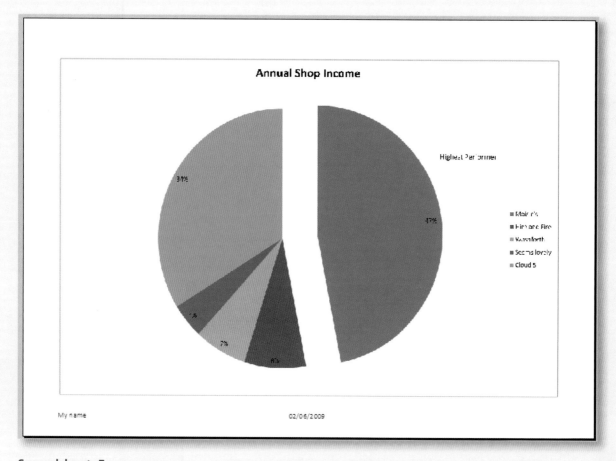

Spreadsheet 5

12 Close all open files.

Creating and using a database

This unit covers the creation, maintenance and interrogation of databases. You will develop a full understanding of the application and learn how to search for records using multiple criteria and create and format a range of reports.

At the end of the unit you will be able to:

➡ create a database file, set up fields and enter data

➡ import a data file

➡ update and interrogate a database using complex search criteria

➡ create and format database reports.

Using Access

Information is stored in a database under headings or **field names** in the form of **records**. These are normally displayed in a **table**, with each individual record making up a single row. Each data item is a **field**.

Field name

Name	Size of Department	Salary	Date of birth
Bill	34	£25,000	12/10/80
Salina	28	£18,500	14/06/82
Wilma	16	£22,750	11/03/79

Record   Field

You can use a spreadsheet application such as Excel to store records. However, its database facilities are limited. If you use a dedicated relational database such as Access, you can do much more with your records including linking and searching across a number of tables and using various objects to display, edit or interrogate the data. In particular:

● **forms** can be used as an alternative method for entering or editing records

● **queries** are designed to help you search for records based on specific criteria

● **reports** allow you to display the records in different ways.

Creating a database file

An Access file will contain all the objects you create for storing and displaying your records, but it has to be created first, before you can enter any data.

create a file

1 Click on a shortcut icon or go to **Start – All Programs – (Microsoft Office) – Access**.

2 On the opening **Getting Started** page, click on **New Blank Database**. (Make sure your window is maximised or you may not see this.)

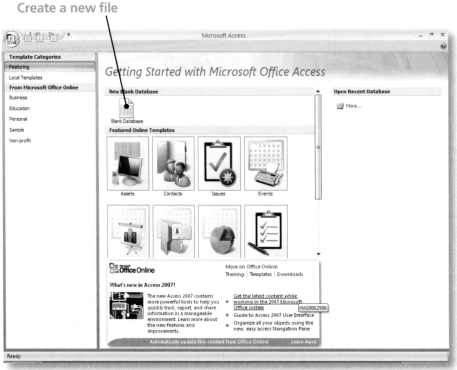

Fig. 3.1 Getting Started

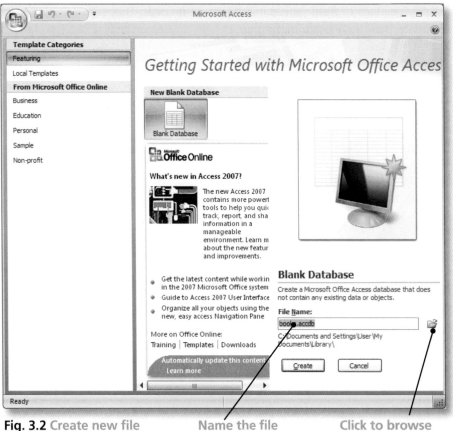

3 When the **Blank Database** pane opens, click in the box and give the file a suitable name. The extension **.accdb** will be added automatically to show it is an Access 2007 database.

4 To change the location for storing the file, click on **More** and browse for a suitable directory or folder. At this stage you could create a new folder to hold the database.

5 Click on **Create** and the new file will open.

Fig. 3.2 Create new file   Name the file   Click to browse for folder storage location

Creating copies

Once a database file has been created, you can make copies of objects within it or copy the entire database. In particular make sure you copy important files, and name the copies carefully, so that you have a backup copy in case the original is lost or corrupted.

save a copy of the file

1 Click on the **Office** button.

2 Hover over **Save As**.

3 Select the appropriate option such as **Access 2007 Database**.

4 When the **Save As** window opens, you will see that the same database name is displayed with a number 1 added, to show it is a second version.

5 Keep or amend the name and select an appropriate location for your copied file.

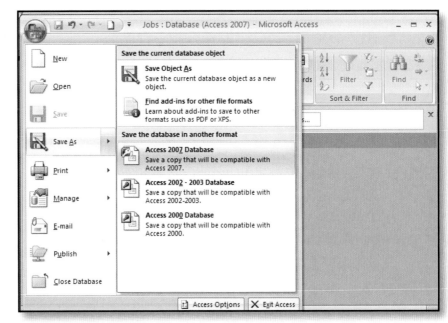

Fig. 3.3 Save copy

Data Types

The data stored under each field name in a database table can be of only one type. This is because Access can only perform calculations, sorts or searches based on recognised data types.

For example, a table may contain the following records:

Name	Size of department	Salary	Date of birth
Bill	34	£25,000	12/10/80
Salina	28	£18,500	14/06/82
Wilma	16	£22,750	11/03/79

The data types will normally be set as follows:

- names will be **Text**
- size of department will be **Number**
- salaries will be **Currency**
- date of birth will be **Date**.

Text – is the data type selected for entries such as Names or Addresses, or for any mixture of letters and numbers (alphanumeric entries) such as Postcodes, Stock codes or Emails that may include figures not used in calculations. The length of any entry is limited to 255 characters.

Number – is applied to numbers such as Ages, Items sold, Measurements or Sizes.

Currency – is used for financial entries such as Costs and Prices that usually display appropriate currency symbols such as £ or $. The number of decimals can be set at 0 (integer), 1 or 2.

Date/Time – is the data type for Dates and Times where you want to be able to perform calculations such as finding an earlier or later day in the month.

Memo – can be selected if you want long textual entries such as Descriptions or Notes.

Yes/No (Logical or Boolean) – is used where a simple answer such as True or False or Yes or No is all that is required.

AutoNumber – is applied when you want to count each record and assign the next consecutive number automatically. This data type will not allow changes.

Primary key

For any professionally designed database, it is important to be able to identify each record, for example to distinguish between people with similar names or items with similar functions. You will find that Access 2007 automatically creates an ID field for this purpose. To make sure that every record has a unique identifier, this field is assigned a **primary key** which means an error message will appear if you try to enter duplicate data. The primary key also plays an important role when you create relationships between different tables.

You can assign a primary key to any field, and suitable fields that will not contain duplicated data include ISBN book numbers, stock codes or National Insurance numbers.

Creating tables

A database file can hold any number of different tables, but these should all have some relationship to one another. For example, a **School** database file could contain tables related to *Staff, Pupils, Exam Results* and *Attendance*.

You can view a table in two different ways: by displaying the records (**Open** or **Datasheet view**) or viewing the underlying structure (**Design view**).

When your newly created file opens, the ribbon will show the **Table Tools** commands related to a table of data temporarily labelled Table1. This has been set up ready for you to customise. It will contain a single field named **ID** with an **AutoNumber** data type. A primary key will have been assigned to it.

You will be in **Datasheet view** ready to add or search through the records contained in the table. Before you can enter any records you must set up the design of your table properly and to do this you need to be in **Design view**.

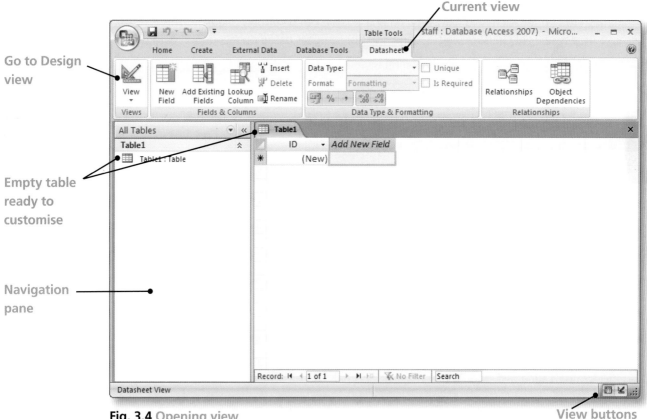

Fig. 3.4 Opening view

There are three important decisions to make when designing your own tables:

- appropriate headings (field names) under which the data will be stored
- the type of data that will be stored in each field
- a name for the table that will indicate the type of information that it contains.

design a table

1 Click on the **Design** button on the ribbon or in the bottom corner of the screen to move to **Design view**.

2 A **Save As** box will appear and you must enter a name for your table. Choose one that relates to the information the table will contain but that is *not* the same as the database filename.

3 In **Design view** you will see the tab on the left displaying the table name and three columns in the main window labelled **Field Name**, **Data Type** and **Description**.

4 Click in the first empty cell below ID in the **Field Name** column and enter the name for your first field, for example **First name**. (You may find some field names such as **Name** or **Date** will be rejected, in which case simply find an alternative.)

Fig. 3.5 Name table

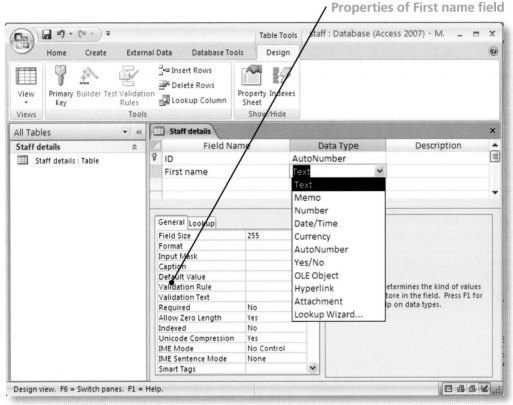

Fig. 3.6 Add field names

5 Press the **Tab** key or click in the next column and select a data type. Text will appear automatically, but click on the drop-down arrow to select an alternative. Note that if you click in the cell and type in the first letter of the data type such as *N* for number or *C* for currency the correct entry will appear.

6 The **Description** column is available to you as the table designer, in case there are any comments you wish to include about a particular field.

7 Each time you add a field name and data type, the properties of that field will be displayed below.

8 To continue designing the table, click in the next empty **Field Name** cell and type in the next field name, for example **Surname**. Select a data type and continue adding fields until all the field names have been added.

9 When your table is complete, click on the **Datasheet view** button to start entering records. You will first have to click on **Yes** to save the changes you have made to the table.

Fig. 3.7 Warning to save table details

Closing

You can have a database file open with any tables that it contains closed, or with a selected table open on screen. To open a table, or other object showing in the navigation pane, double click on its name. Any open object will have a tab displaying its name and an icon to show the type of object it is.

close a table

1 Click on the table **Close** button showing a cross.

or

2 Right click on the named tab in the main window and select **Close**.

close the file but keep Access open

1 Click on the **Office** button and then click on **Close Database**.

exit Access

1 Click on the database file **Close** button.

or

2 Click on the **Office** button and click on **Exit Access**.

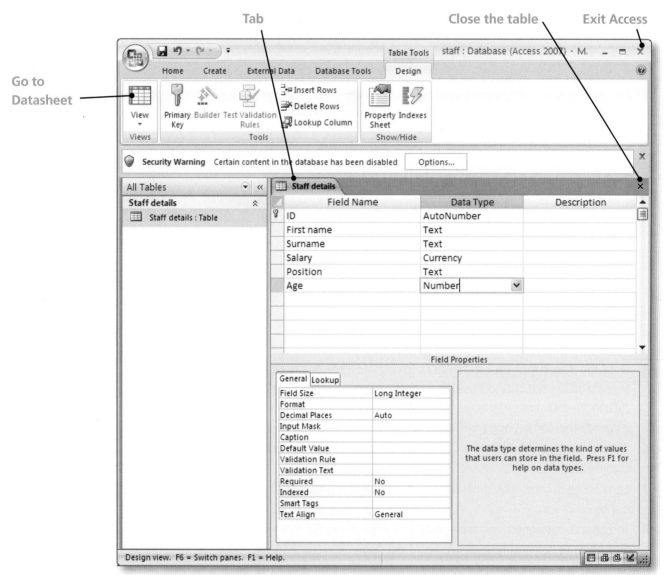

Fig. 3.8 All field names

Check your understanding 1

1 Create a new database file named *Holidays*.

2 Design a table named *England*.

3 Add the following field names:

Field name	Data type
Area	Text
Town	Text
Visits	Number
Recommended	Yes/No

4 Save and close the table.

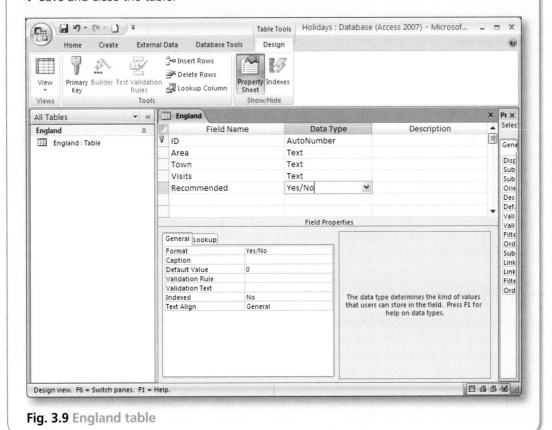

Fig. 3.9 England table

Designing a table in Datasheet view

For more sophisticated, business-orientated databases you can use the ready-made list of field names provided by Access and design a table in **Datasheet view**.

design a table using the field list

1 After creating a new file, click in the first **Add New Field** column.

2 Now click on the **New Field** button.

3 Scroll down the list of **Field Templates** offered and select an appropriate name.

4 Drag it up to the table or double click so that it replaces the *Add New Field* name.

5 Appropriate data types such as currency or number will have been set automatically for each field name listed. If incorrect, click in the **Data Type** box and select an alternative.

6 Continue to work through the list to add further field names.

7   Click on the **Save** button to name and save the table.

8   Records can now be added to the table.

9   Add new fields at any time by reopening the **Field Templates** list and selecting a new name.

Click for alternative data type

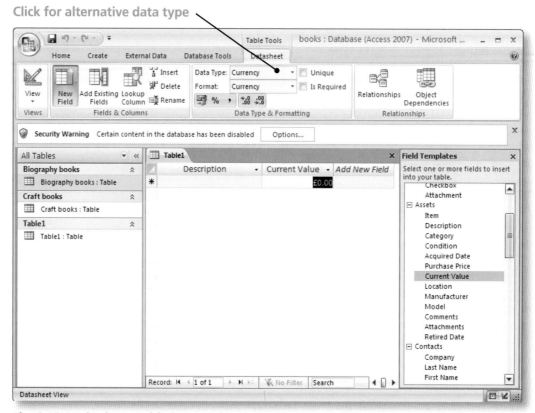

Fig. 3.10 Designing a table in Datasheet view

Entering records

Once the basic table structure has been established, you can add information to your tables in the same way that you enter data into an Excel spreadsheet or Word table. Click in a cell and start typing. Move across the row to the next cell by pressing the **Tab** key or clicking with the mouse.

Note that databases consist of records – you should *not* try to enter data down columns but always complete one full row at a time. You will not be able to add records between those already present. Instead, add a new record at the bottom of the table where an asterisk * is displayed and change the record order by carrying out a sort.

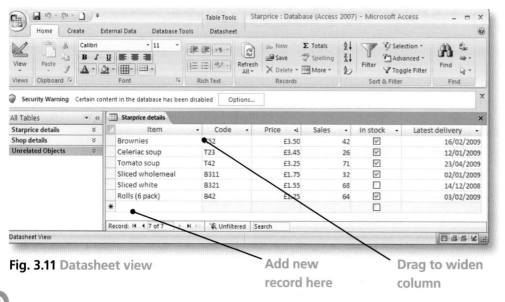

Fig. 3.11 Datasheet view

Add new record here

Drag to widen column

display data in full

1   Position the pointer on the boundary between field names.

2   Hold down the mouse and drag the boundary in or out.

3   You can also double click the mouse to set the column width to fit the longest entry.

Check your understanding 2

1 Open the *Holidays* file you created earlier.

2 Open the *England* table in Datasheet view.

3 Enter the following records:

Area	Town	Visits	Recommended
Southwest	Torquay	3	Yes
Southeast	Brighton	2	No
South	Bournemouth	4	Yes
South	Portsmouth	1	No

4 Make sure the data is displayed in full.

5 Now change *Portsmouth* to *Southsea*.

6 Save and close the file.

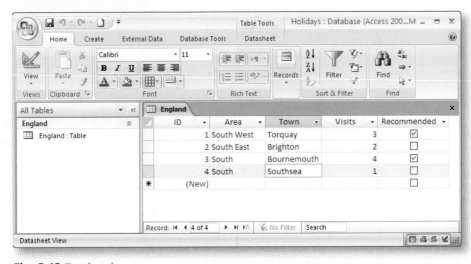

Fig. 3.12 England

Deleting records

A database table is made up of **columns** (fields) and **rows** (records). As with an Excel spreadsheet, it is easy to delete either of these by selecting it and then using one of the **Delete** options.

Take care when deleting records as, once confirmed, you cannot undo the action.

delete a record

1 In **Datasheet view**, click on the row selector to the left of the first field or click on and drag to select more than one adjacent record at the same time.

2 Press the **Delete** key.

or

3 Right click and select **Delete Record**.

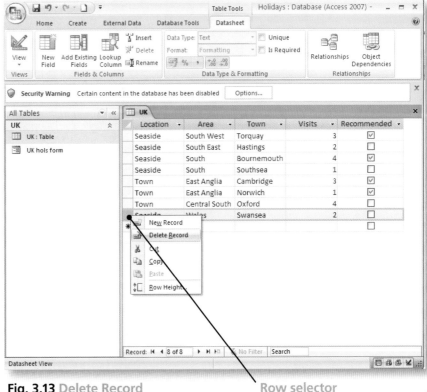

Fig. 3.13 Delete Record   Row selector

Fig. 3.14 Delete warning

4 Click on **Yes** when the warning message appears.

Fig. 3.15 Open recent database

Opening a saved database file

Once a database file has been created and saved on your computer, you can open it in one of two ways. Either locate the file from the desktop by opening a folder in which it is stored and then double click on this to open the file directly, or open Access and then browse for the file.

open a file in Access

1 Click on the **Office** button.

2 Database files you have already created and opened recently will be listed in the **Recent documents** window. Click on one to open.

3 Otherwise, click on the **Open** button to open the dialog box.

4 Navigate to the folder containing the database file.

5 Select its name and click on **Open**.

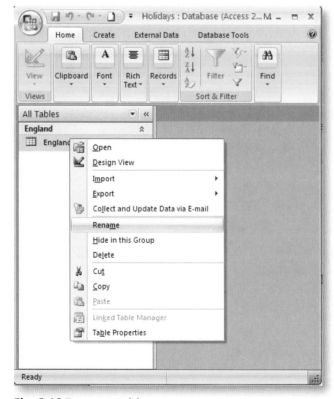

Fig. 3.16 Rename table

Renaming tables

If at some stage you decide to rename any tables or other objects you have created, you can do so as long as they are closed.

rename a table

1 Right click on the object name in the navigation pane.

2 Select **Rename**.

3 Type a new name for the object.

Check your understanding 3

1 Open the database file *Records* provided on the CD-ROM accompanying this book.

2 Rename the table named *Pop music* so that it is now named *Charts*.

3 Save and close the file.

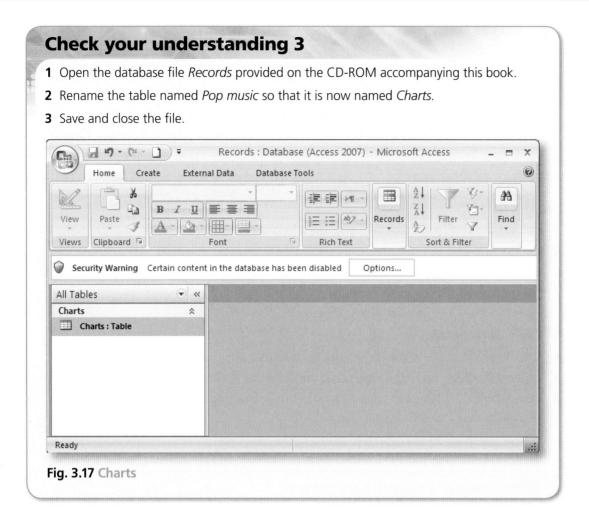

Fig. 3.17 Charts

Adding or deleting fields

If you do not want to retain the **ID** field, or if you change your mind about other fields, you can delete them easily when designing your table. If they have a primary key defined, you will need to click on **Yes** when the warning message appears. Once you have entered any records, you will always be warned that a deletion cannot be reversed.

You can also add fields to your table if you want to do so after it has been created, or change their order across the table.

delete fields

1 Open the table in **Design view**.

2 Click on the pale blue square next to the field name when the pointer shows a black arrow. The selected row will display an orange border.

3 Click on the **Delete Rows** button and confirm the deletion when a warning message appears.

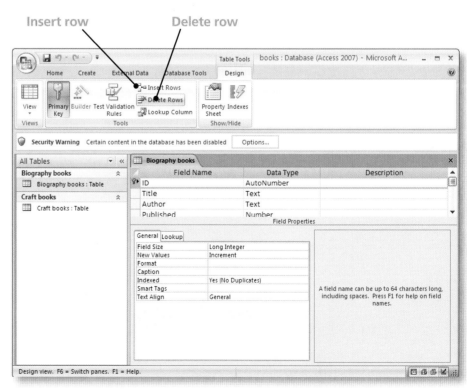

Fig. 3.18 Delete field row

add a field

1 Click in the next empty row to add a field at the end of the list.
2 To insert a field, click to select the row *below* the position for the new field.
3 Click on the **Insert Rows** button.
4 A new row will appear and you can now enter the field name and data type.
   or
5 View the table in **Datasheet view** and type in the empty field column to the far right where it says *New Field*.

reorder fields

1 In **Design view**, select the field you want to move.
2 Click on the blue selector square and hold down the mouse button.
3 Drag the field up or down the list. Its new position will be marked by a black horizontal line.
4 Let go when it is in the correct place.
   or
5 Go to **Datasheet view**.
6 Click to select the field column you want to move.
7 Move the pointer up to the field name and hold down the mouse button.
8 Drag the field horizontally across the table until it reaches the correct position. This will be marked by a black vertical line.
9 Let go and the field will drop into place.

Check your understanding 4

1 Open the database file *Holidays*.
2 Add a new field between *Town* and *Visits* that has the name *Location* and a Text data type.
3 Remove the ID field.
4 Move the field *Visits* to below *Recommended*.
5 Save these changes and close the table.
6 Finally, rename the table *UK*.

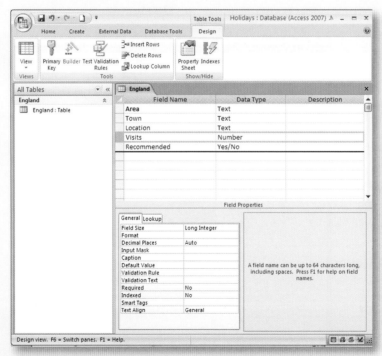

Fig. 3.19 UK table – Step 3

Field properties

When you start typing in details for a new field during the design of a table, a window labelled **Field Properties** opens up below it. Here you will find the settings for each data type which you can accept or amend. (Bear in mind that if records have already been entered, changes to field properties may not affect them all. For example, figures after a decimal may not be displayed when changing from integer/whole number format.)

The major properties include:

- **Field size** – this is the space left for characters you type in, or sets the number type
- **Format** – this sets the appearance (e.g. currency or dates)
- **Input mask** – you can use this to limit how users type in their entries
- **Caption** – this is used where the field name is not wanted on display
- **Default value** – this can help speed up entry of repeated data
- **Validation rule** – this will help prevent ineligible or unwanted entries
- **Validation text** – the error message displayed if a validation rule is broken
- **Required** – this prevents a field being left blank
- **Allow zero length** – this will allow blanks
- **Indexed** – you can set an index to speed up searches.

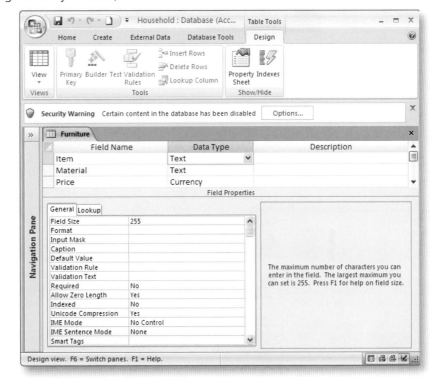

Fig. 3.20 Field Properties

For the assessment, make sure that your computer is set to show the appropriate date – for example, for UK, the day precedes the month. If you need to change your computer settings, the **Regional** and **Language Options** are accessed from the **Control Panel**.

The default settings for the main data types include:

Data type	Text	Number	Currency	Date/Time
Field size	Limited to 255 characters which you can reduce to save space	The default is a Long Integer, i.e. no decimal places displayed		
Format		You will need to set this to Double to display any decimals	It is set to display £ but you can change this to $ or no symbol. Fixed, unlike Standard, does not show comma separators	Usually set as long time (e.g. 5:34:23 pm) or short date (e.g. 12/11/2008). You can select long or medium (e.g. 12-Nov-2008)
Decimal places		Usually none set so you can select the required number	Usually set at 2 but you can select none or 1	

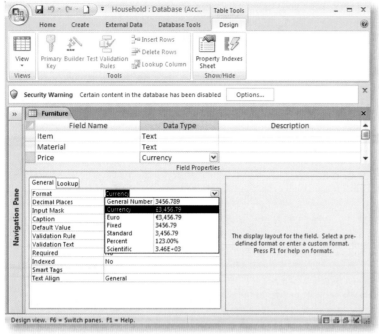

Fig. 3.21 Set properties

change Field Properties

1   In **Design view**, click on the field whose properties you want to change.

2   In the **Field Properties** box, click in the appropriate cell to select from a drop-down list of alternative entries or type in an entry. (Note that a drop-down arrow may not be displayed automatically but will appear when you click in the cell.)

3   Repeat with any other fields.

4   Save the table before returning to **Datasheet view** or closing the file.

    or

5   In **Datasheet view**, click on the **Format** box below the **Data type** box to select a different number, currency or date format. There are also shortcut buttons to increase or decrease decimal numbers, add percentage or currency symbols and include a thousand separator.

Number formats

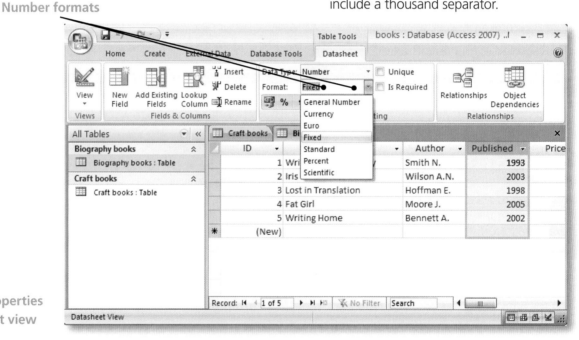

Fig. 3.22 Properties in Datasheet view

?????

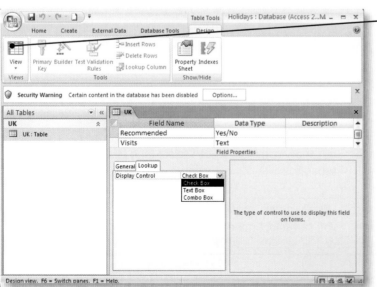

set the Yes/No display

1   In the **Properties** box, click on the **Lookup** tab.

2   To display a checkbox in the table, select **Check Box**. When you enter records, you will need to click to place a tick in the box for **Yes** and leave blank for **No**.

3   For a normal cell where you type in the words Yes or No, select **Text Box**.

Fig. 3.23 Yes/No box

Checking properties

It is important to check that the formats you have set are appropriate, so you need to be able to switch to **Datasheet view**. There are various ways to view records, or the fields ready for record insertion:

- Click on the **Datasheet view** button when in **Design view**.

  or

- Press the function key **F6** to move to that view.

- Open a database file and then double click on the name of the table showing under **All Tables** in the navigation pane.

Check your understanding 5

1 Open the database file *Leisure Centre* provided on the CD-ROM accompanying this book.

2 Open the *Staff* table and look carefully at the records. Note that *Rate per hour* shows as an integer (whole number) with no £ symbol and the *Date started* format is a short date.

3 In **Design view**, make the following changes:

- **a** Display *Rate per hour* as currency with a £ symbol and 2 decimal places.
- **b** Change the date format to a long date – for example, 21 June 1999.
- **c** Change the Yes/No checkbox to text.
- **d** Remove the ID field.

4 Switch to Datasheet view and check that these changes have been made.

5 Save and close the file.

Fig. 3.24 Change properties

Validation

To make sure people working with a table only enter appropriate data, you can restrict what is entered by enforcing a **validation rule**. If this is broken, an error message – the **validation text** – will appear offering guidance on how to complete the entry correctly. For example, if your recipes must all be cooked in less than two hours, you can display an error message if anyone tries to enter a cooking time for a recipe that is over 120 minutes.

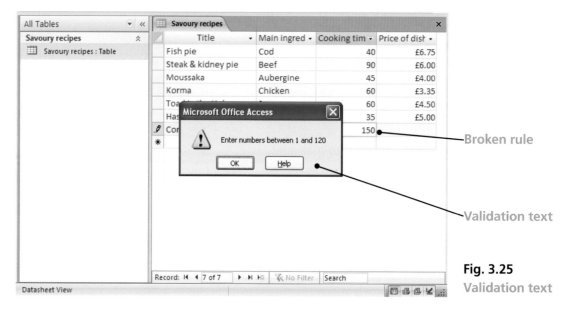

Fig. 3.25
Validation text

Validation rules must use acceptable **expressions** so that they can be interpreted by Access. These include the following.

Text

- For a single entry or range of text entries, type the allowable text between inverted commas (quote marks: " "). For example, to allow only London you would enter **"London"**.
- To limit entries to London, Paris or Rome, you would enter **"London" Or "Paris" Or "Rome"**.
- To reject an entry, use **<>**. For example, you would enter **<>London** to reject the entry London.

Numbers and Dates

(Note that these should not include symbols or text, for example £ or st.)

- For a number larger than or date after... use **>**. For example, **>0** will prevent entries that are negative or zero.
- For a number less than or a date before... use **<**. For example, **< 21/3/06** would mean only UK dates earlier than 21 March 2006.
- For equal to or less than... use **<=**. For example, **<=5** (it cannot be more than 5).
- For equal to or more than... use **>=**. For example, **>= 50** (it must be 50 or above).
- Between two numbers use **> And <**. For example, **>3 And <10** (any number between 3 and 10).
- For not equal to... use **<>**. For example, **<>20** (not 20).

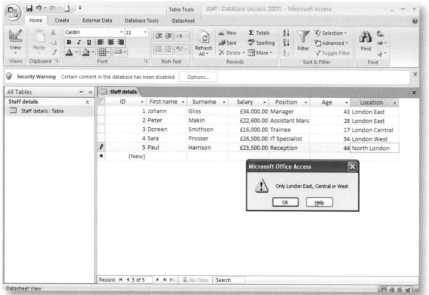

Fig. 3.26 Breaking validation rule

set validation rules

1 In **Design view**, click on the field you want to set the rule for.

2 Click on the **Validation Rule** property box and type in the expression.

3 Click in the **Validation Text** property box and type in the error message users will see if their entry breaks the rule.

4 Return to **Datasheet view** and test that the rule works.

5 When it does, you must click on **OK** and type in a new entry that does not break the rule.

Check your understanding 6

1 Open the database file *Chemists* provided on the CD-ROM accompanying this book.

2 Open the *Luxury Products* table in **Design view**.

3 Set a validation rule for the *Price* field that will only allow prices between £20 and £40.

4 Add appropriate validation text.

5 Change to **Datasheet view** and try to enter the following record:

   Sunflower bath, 12, £42.50, gift set

6 Change the price to £35.50.

7 Save and close the file.

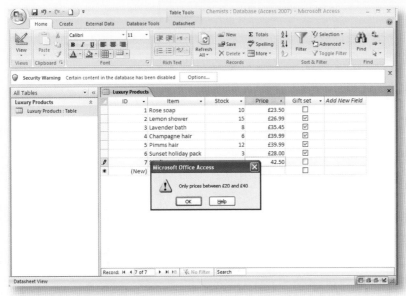

Fig. 3.27 Validate rule set

Printing tables

When printing a table, you will see that the table name, the date and page numbers are added automatically in the top and bottom margins, and the default orientation is Portrait. As you can have a number of field columns in a table, this often means that part of the table extends onto a second or even third page. To make sure you display the records clearly, you should check the table first in **Print Preview**.

preview a table

1 Click on the **Office** button.

2 Hover over **Print** and click on **Print Preview**.

3 Click on the **Two Pages** button to check whether the table extends across more than one page.

4 If necessary, click on the **Landscape** orientation button to bring all the fields onto a single page.

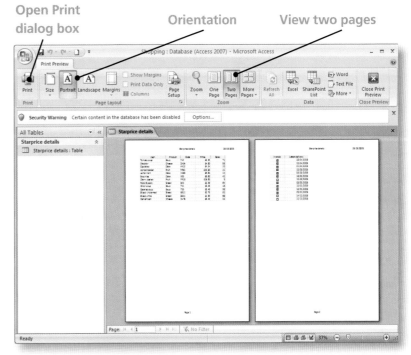

Open Print dialog box

Orientation

View two pages

Fig. 3.28 Print Preview table

print a table

1 Go to the **Office** button – **Print** – **Quick Print** to print the table using the default settings.

2 Preview the table and make changes before clicking on the **Print** button to open the **Print** dialog box.

or

3 Go to the **Office** button – **Print** – **Print**.

4 Make changes to the number of copies or pages to print before clicking on **OK**.

5 If you select a limited number of records before opening the **Print** box, you can choose to print just that selection.

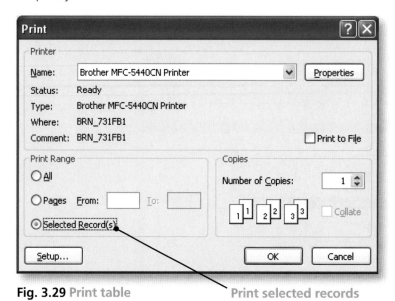

Fig. 3.29 Print table   Print selected records

Check your understanding 7

1 Open the file *Shopping* provided on the CD-ROM accompanying this book.

2 Print one copy of the *Starprice details* table making sure it fits on one page.

3 Now print one copy of just the first four records.

4 Save and close the file.

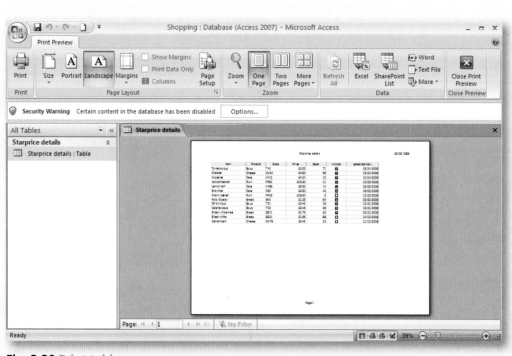

Fig. 3.30 Print table on one page

Forms

As well as entering data into a database table where all the records are visible at the same time, you can use forms that will enable you to view and work with one record at a time. This is often a far easier method for keying in data.

create a form for data entry

1 On the **Create** tab, click on **Form**.

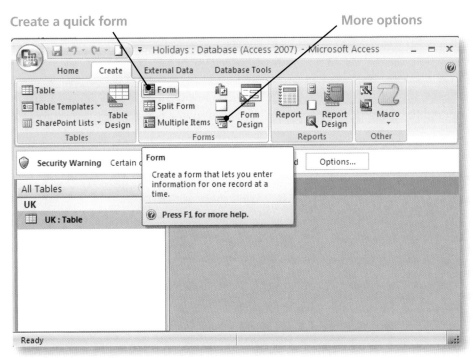

Fig. 3.31 Create form

2 A form will appear on a new tab, displaying the first record in the table. It will be in layout view.

3 You can apply a different style of layout to your form by choosing from the **AutoFormat** gallery of styles.

4 If there is more than one table in the database, you will need to use the **Form Wizard** in order to select the appropriate data on which to base the form.

   a Click on **More Forms – Form Wizard** to start the process.

   b Select the table from the drop-down list.

   c Add the fields you want to include in the form. For all fields, click on the double arrow.

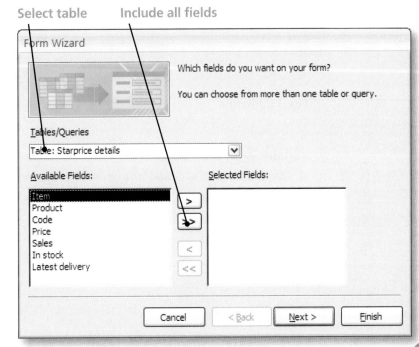

Fig. 3.32 Form Wizard

d Either work through to select layout and style or just click on **Finish** and the form will appear.

5 To enter records, change to **Form view** by selecting the option from the drop-down arrow on the **View** button on the **Home** tab or from the **View** buttons in the bottom right-hand corner.

6 Navigate between records by clicking on a **Record** selector button at the bottom of the form. You can move to the first, last, first empty or next/previous record.

7 You can also type the record number in the box and press **Enter** to move to a specific record.

8 When you want to enter data in one record, move between fields by pressing the **Tab** key.

9 Click on any field to edit the data.

10 To name and save the form, right click on the **Tab** and select **Save** or click on the form **Close** button and the **Save** box will appear. The default name is the same as the table name.

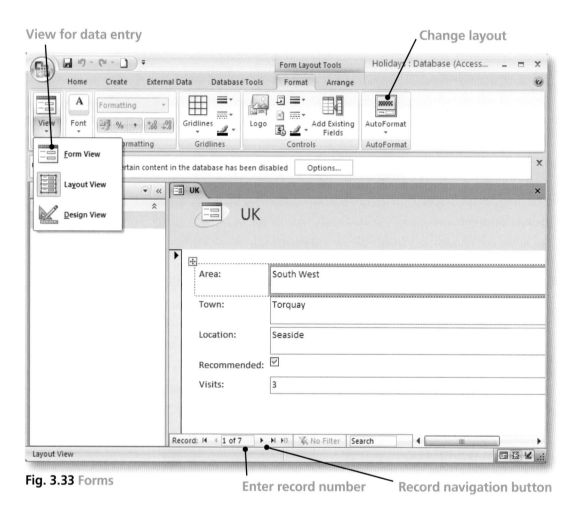

Fig. 3.33 Forms

rename a form

1 Right click on the closed form in the navigation pane.
2 Select **Rename**.
3 Type in a new name.

Check your understanding 8

1 Open the database file *Flowers* provided on the CD-ROM accompanying this book.

2 Create a form based on the table.

3 Navigate to the first empty record and add the following:

Aconite, Yellow, 15, £0.45, September

4 Now find the record relating to *Pink Hyacinth* and change the *Number in stock* to 56.

5 Save the form as *Bulb Form* and close the file.

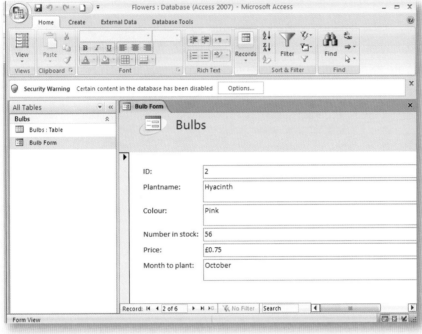

Fig. 3.34 Bulb form

Printing forms

The default print setting for a quick form is Portrait orientation with all the records displayed one below the other on the page. To print one record, display it on screen and then click on **Selected Record** in the **Print** dialog box.

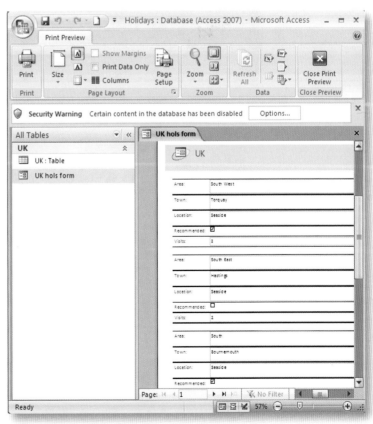

Fig. 3.35 Print form

Sorting

Although you cannot enter a record between others in a table, you can reorder them by carrying out a sort. This can be in either **ascending** or **descending** order:

- **Ascending** – alphabetical from A to Z, lowest to highest number, or earliest to latest date or time
- **Descending** – from Z to A, highest to lowest number, or most recent to earliest date or time.

As Access does not allow records to be split, you can select the field on which to base the sort and all the data for each record will keep its integrity when the sort is carried out.

sort records in a table

1 Open the table.

2 Click on any entry in the field on which to base the sort.

3 Click on the **A–Z** button for an ascending sort, and the **Z–A** button for a descending sort.

4 The sort options are also available from the drop-down arrow in each **Field Name** box.

Descending order

Sorting on price

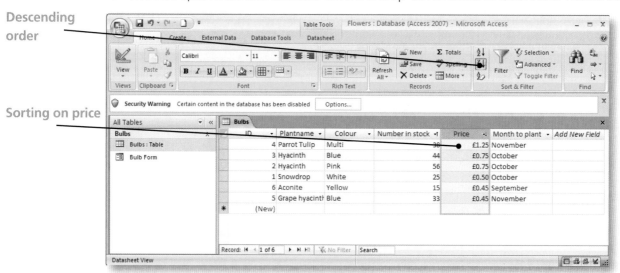

Fig. 3.36 Sort

Check your understanding 9

1 Open the database *Shopping*.

2 Sort the *Starprice details* table alphabetically by *Product*.

3 Now sort the records in descending order of *Latest delivery*.

4 Save and close the file.

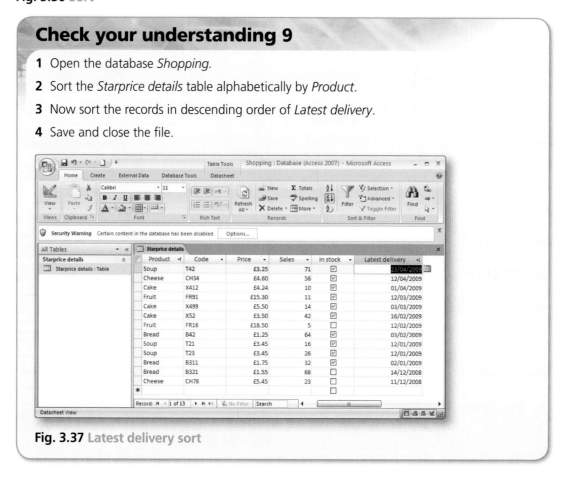

Fig. 3.37 Latest delivery sort

Searching a database

There are three methods for locating records that meet certain criteria:

- using the **Find** tool
- applying **filters**
- creating **queries**.

Find will locate part or all of a field such as a single price, size or colour. You need to move along the row to see other data relating to the selected record. If you want to replace an entry, find it first or use the **Replace** tool for multiple entries.

Filters allow you to temporarily hide records in a table that do not match your criteria, so that you are left with all the data related to the records that do match. Thus you can display all records for items that are the same size, price or colour. You can also carry out logical searches. For example, you could display all the records with a start date between a range of dates or where ages are above or below a particular figure.

Although you can print out the subset of the table displayed after applying a filter, filters are only temporary.

Queries are the only permanent way to search your records. You can design a query to display a limited number of fields in any order, and the query can be named and saved so that it can be run in the future.

Find and replace

In Access, when using **Find** to search a table, searching is normally restricted to a single field – the one marked by your cursor. If this is not the correct field to search, you can:

- move the cursor to the correct field before starting a search.
  or
- select the option to search the entire table.

find data

1 Open the table you want to search.
2 On the **Home** tab, click on **Find**.
3 In the **search** box, enter the exact data you want to locate.
4 In the **Look In:** box, check that the correct field will be searched or change to the table name.
5 In the **Match:** box, select an option such as *Any Part of Field* if the entry includes other characters such as currency symbols.
6 Click in the **Match Case** checkbox if you want the search to be case sensitive.
7 Click on **Find Next** to move to the first matching entry.
8 Keep clicking on the button to work through the table.

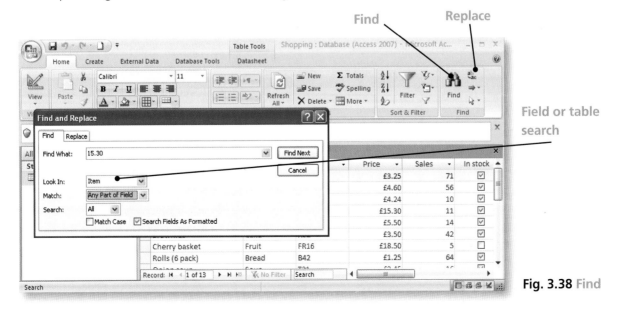

Fig. 3.38 Find

replace entries

1 Open the **Find and Replace** box by clicking on **Replace** on the **Home** tab or click on the labelled tab if the box is already open.

2 Enter the data you wish to find in the **Find what:** box.

3 Enter the replacement entry in the **Replace with:** box.

4 Click on **Replace All** if you are sure it will be an accurate replacement.

5 Otherwise, click on **Find Next** to locate the entry.

6 Click on **Replace** to replace it or **Find Next** to skip to the next entry.

Check your understanding 10

1 Open the database file *Shopping*.

2 In the *Starprice details* table, replace any entries for 2008 with 2009.

3 Now find which product had sales of 64 items.

4 Close the file.

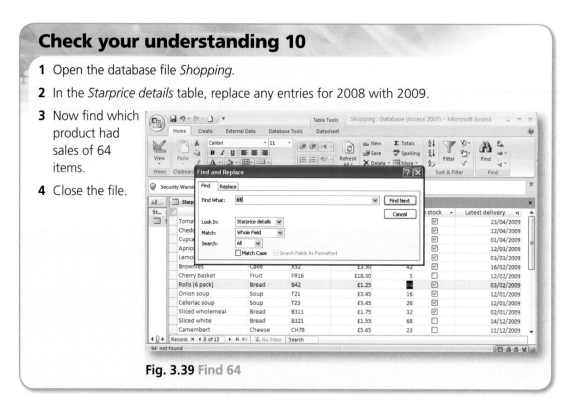

Fig. 3.39 Find 64

Filters

If you are filtering on one criterion, you can use **Filter by Selection**. For several criteria, you need to use **Filter by Form**.

Having carried out one filter, you can carry out another to search that subset of records or take off the filter to show all the records again.

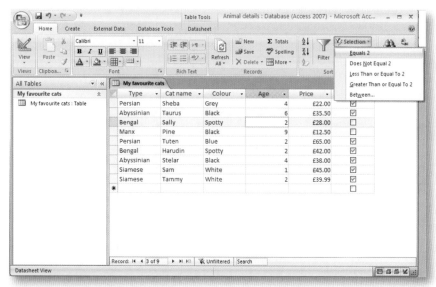

Fig. 3.40 Selection filter

apply a selection filter

1 Click on one example of an entry that matches your criterion.

2 Click on the **Selection** button in the **Sort & Filter** group.

3 Select from the options available. For example, you can look for records containing or not containing selected text, or that equal or are more or less than a particular number.

4 For figures such as prices or dates, click on **Between** and you will be offered a box and can type in which range to search.

5 All records matching the selected criteria will be displayed and the rest of the table will be hidden.

6 You can now print out the records.

7 To remove the filter and view the entire table, click on the **Toggle Filter** button.

8 You can also click on the **Filter** button for a range of sorting options and to clear the filter.

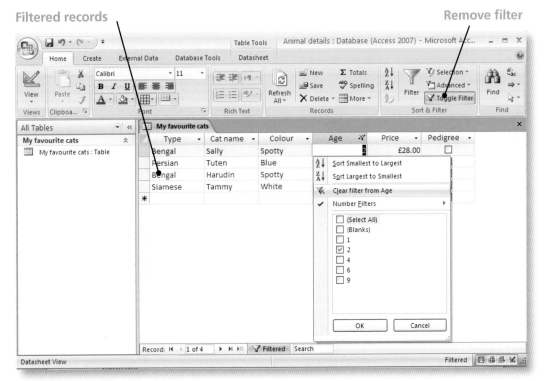

Fig. 3.41 Clear filter

filter by form

1 In the **Sort & Filter** section, click on **Advanced – Filter by Form**.

2 You will see all your field names and an empty row for your entries.

3 For each criterion, click in the relevant field box and either select an entry from the drop-down list or type in an expression such as > (greater than) or <> (not equal to). (See page 146 'Validation' for a list of acceptable expressions.)

4 Check the form carefully – for example, make sure a previous criterion has not been set by mistake – and then click on **Toggle Filter**.

5 Matching records should be found.

6 Remove the filter by clicking on the **Toggle** button again.

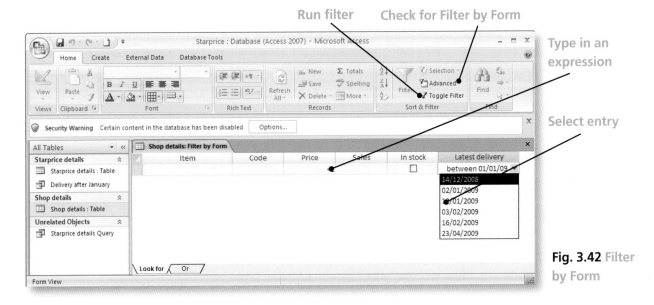

Fig. 3.42 Filter by Form

Check your understanding 11

1 Open the database file *School* provided on the CD-ROM accompanying this book.

2 In the *Summer school* table, apply a filter to display only English classes.

3 Print a copy of the records making sure it fits onto one page.

4 Now find out if any Y8 classes have over 12 students.

5 Close the file.

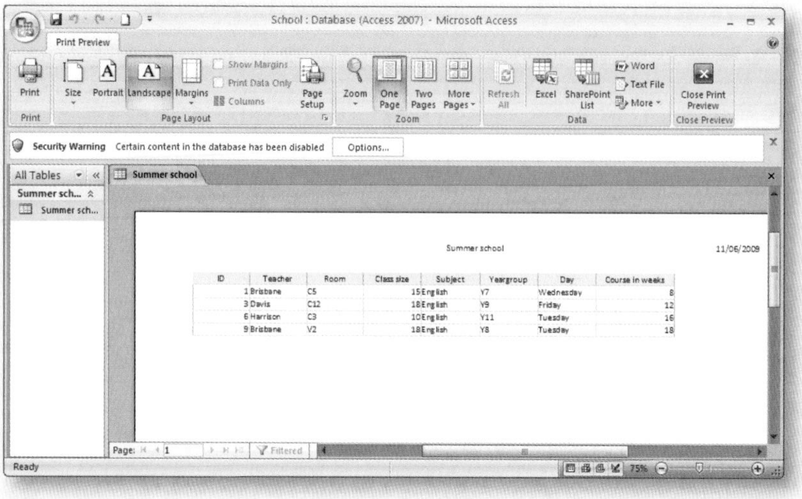

Fig. 3.43 Summer school filter

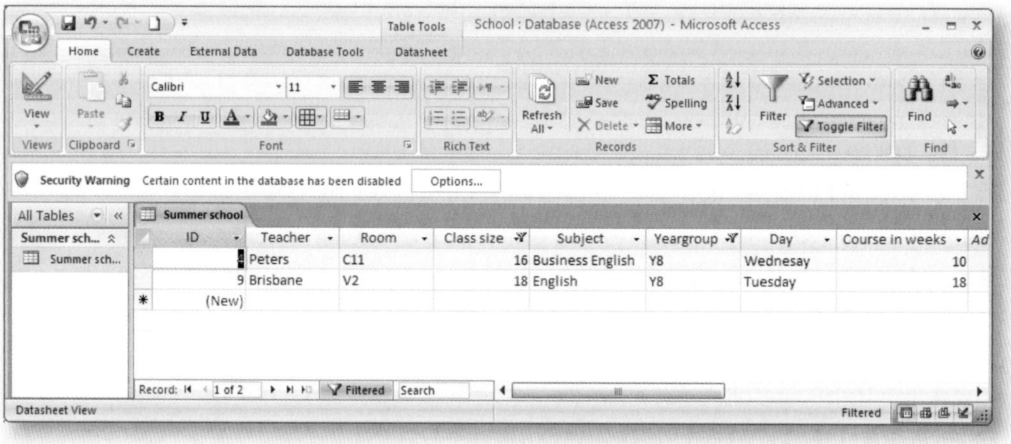

Fig. 3.44 Y8 classes

Queries

The main method used to search a database for records meeting various criteria is by designing and running select queries. You can use a wizard to help you, but it is easier to design a simple query using the grid method.

The main decisions that need to be taken when creating queries include:

● which fields to display and in what order

● what criteria to use when searching

● how to sort the resultant records

● whether to build in calculations such as totals or averages.

create a query

1 On the **Create** tab, click on **Query Design**.

2 A **Show Table** window will appear. Select the table you want to search and click on **Add** to add it to the top of the grid. Close the **Show Table** window.

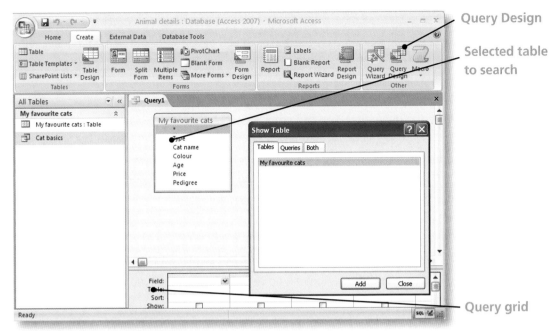

Query Design

Selected table
to search

Query grid

Fig. 3.45 Show table window

3 The table will display a list of all the fields it contains. You can now select which fields to be displayed each time you run your query. There are three ways to do this:

- Double click on a field name to add it to the grid.
- Click and drag each field name onto the Field row of the grid.
- Click in the Field row cell and select the field name from the drop-down list.

Make sure you add any field that is going to be used as the basis for a search.

4 Note that the table name will be added to the grid automatically. This is useful when searching across a number of different tables.

5 When you have added the fields and set the search criteria as explained below, click on the **Datasheet view** button or the red exclamation mark (**Run** button) to view the selected records.

6 To name and save the query and add it to the navigation pane, click on the **Save** or **Close** button.

7 Return to the **Design view** of the query if you need to make any adjustments.

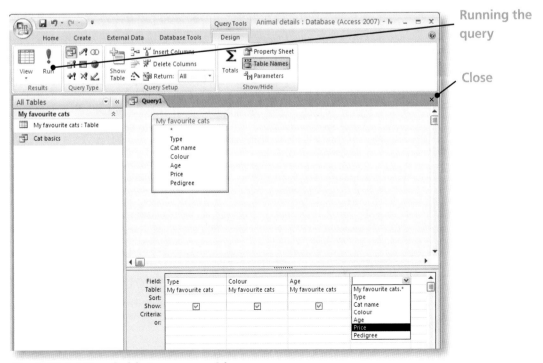

Running the
query

Close

Fig. 3.46 Adding fields to query grid

Printing queries

Queries are printed in exactly the same way as tables. If you have saved and named a query, its name will show at the top of the page, rather than the default temporary name Query1.

Query criteria

For any field added to the query grid, you can enter an expression in the criteria row so that only records meeting those criteria will be located and displayed when the query is run.

1 You can search for an exact match by entering the text, number or date exactly as it is displayed in the table. When you click away from the cell, Access will add:

- quotation marks to show a text data type has been accepted
- hash symbols (##) to show a date or time data type has been accepted
- no extra symbols when a number or currency data type is accepted.

2 You can also use <> or **Not** to locate records that do *not* match an entry.

For example, **<>blue** or **Not blue** will mean no records will be displayed that contain the word *blue* in the chosen field.

3 If you do not know all the characters or figures in the entry, you can use **wildcards** – symbols representing missing parts. When you click away, Access will add the word *Like*.

- An **asterik (*)** will find all the unknown characters. For example:
  - o If you know the name starts with **sta** and ends with **ing** you can enter **sta*ing** and the search will locate entries including *staying, starting* and *starving.*
  - o For missing figures, ***123** will find any numeric entry that ends with these figures and ***123*** will find numbers containing these figures.
- A **question mark (?)** will find a single missing character. For example:
  - o **t?p** will find entries such as *tip, tap* and *top*
  - o **25/?/2009** will find the 25th of every month in 2009.

4 For numeric entries – numbers, currency, date and time – you can use relational criteria such as:

- Greater than >
- Less than <
- Not equal to <>
- More than or equal to >=
- Less than or equal to <=
- Between two numbers or dates – e.g. **Between 1/11/08 and 21/3/09**.

5 You can use logical operators where there is a combination of criteria:

- When both conditions must be met, use **AND**. For example, after one date and before another, you could enter: **>1/11/08 AND <21/3/09**.
- Where either condition can be met, use **OR**. For example, in a location field, you could search for staff working in either **Sweden OR Holland**.
- You could also use the **OR** row to add the second criterion here.

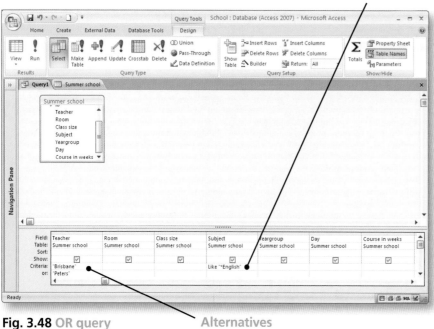

Fig. 3.48 OR query

Check your understanding 12

1 Open the database file *School.*

2 Create a query that will display all the fields in the *Summer school* table.

3 Search for any teachers with names beginning with *B* using a wildcard search.

4 Return to **Design view** and remove this criterion from the grid.

5 Now display records for all classes of 12 or more students taking place in room V2.

6 Save as *V2 class.*

7 Print a copy of the records in Landscape orientation.

8 Close the file.

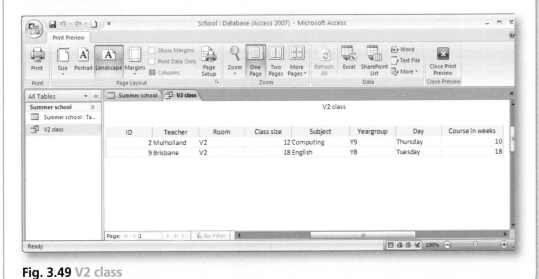

Fig. 3.49 V2 class

Calculated fields

You can add new fields when designing queries that will carry out calculations based on the data in the table. In the **Field** row on the query grid, give the field a new name, followed by a colon and then the formula.

As in Excel, when entering formulas you use the four operators +, -, * and / but you must enter the field names in square brackets. So, for example, to work out a final price by multiplying stock numbers by unit price, your new field name would appear as:

Final Price:[Stock number]*[Unit Price]

When the calculations appear, you may need to amend the properties so that you display the correct number of decimals or a currency symbol.

set calculated field properties

1 Right click on the field name in **Design view**.

2 Select **Properties**.

3 Set the new properties in the window that opens by clicking in the **Format** or **Decimal Places** boxes.

Fig. 3.50 Calculated field properties

159

Check your understanding 13

1 Open the database file *Bookshop* provided on the CD-ROM accompanying this book.

2 You are going to design a query to work out the shop income.

3 Add the following fields to the grid: *Title, Author, Price,* and *Purchased.*

4 Create a calculated field named *Income* that uses the formula Price x Purchased.

5 Run the query.

6 Save the query as *Income*.

7 Print the query making sure it fits on one page.

8 Save and close the file.

Fig. 3.51 Income

Hiding and sorting

There are two extra rows in the query grid that you can use to change the display of the records when a query is run.

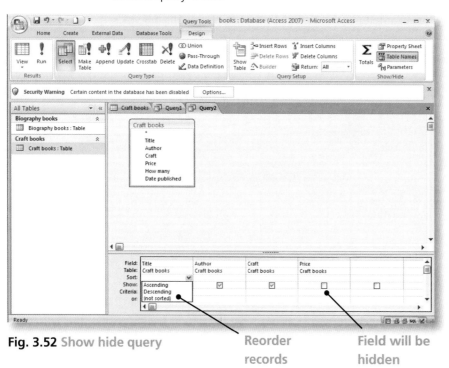

Fig. 3.52 Show hide query

Reorder records

Field will be hidden

- The **Show** row has a tick box. If you remove the tick in any box, that field will be hidden from view. As you must include each search field on the grid so that criteria can be set, you can remove the tick if you do not want the field displayed when the query is run.

- The **Sort** row can be clicked in any field on which you want to base a sort. Select ascending or descending order before running the query.

Check your understanding 14

1 Open *Bookshop* and create a new query to display all the books that cost £12.99 or more.

2 Only display the following fields: *Title, Author, Purchased*.

3 Sort alphabetically by *Title*.

4 Save as *Expensive books*.

5 Print a copy of the query.

6 Close the file.

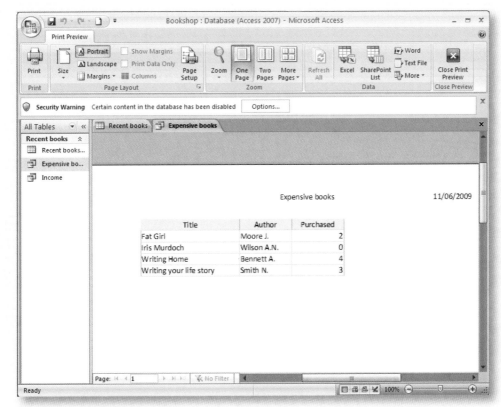

Fig. 3.53 Expensive books

Importing and exporting data

Access allows you to import and export data between a range of applications including other databases, spreadsheets and text or data files such as **CSV (Comma Delimited)** files.

import a data file

1 Open the database file receiving the data.

2 On the **External Data** tab, click on **Text File**.

3 In the dialog box that opens, click on the **Import** option and browse for the file.

Select option

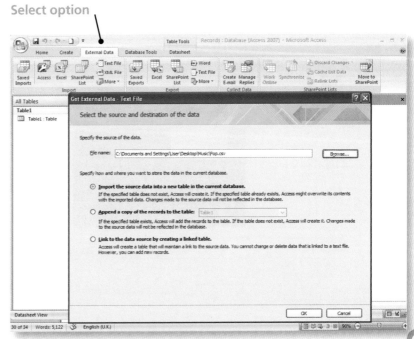

Fig. 3.54 CSV1

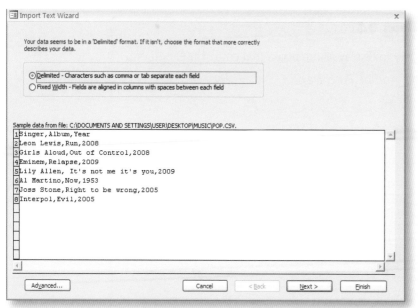

Fig. 3.55 CSV2

4 Click on **OK** and check that the system has picked up the right type of file; work through the wizard by clicking on the **Next** button.

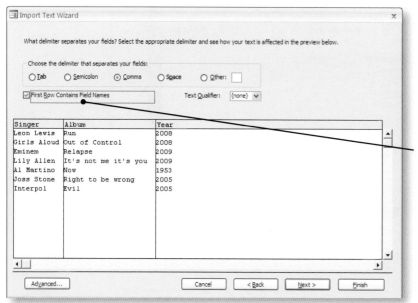

Fig. 3.56 CSV3

5 In the next window, click for headings if they are present as these will form the Field Names in the new table.

Headings identified

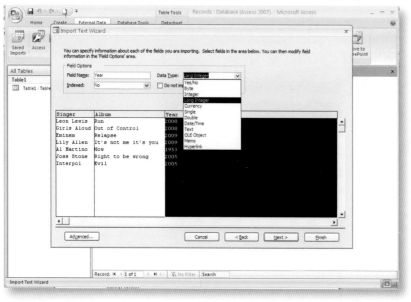

Fig. 3.57 CSV4

6 Now you can click on any field and set the data type, or leave this until the data has been imported.

7 When you click on **Next**, Access will add an ID field with a primary key set. You can click on **No** to remove it or select your own field for the primary key.

8 Finally, enter a name for the table and complete the import.

9 Back in your database, the table will have appeared.

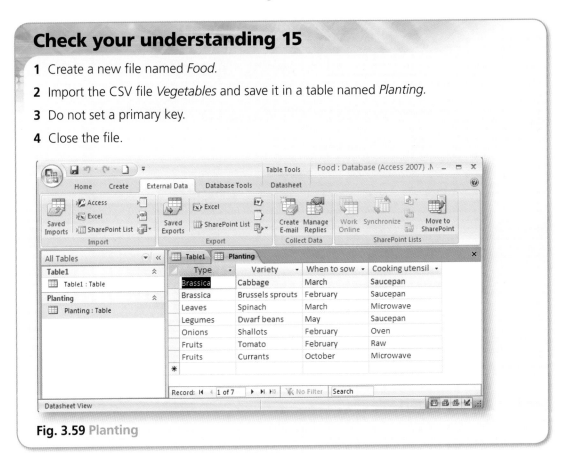

Imported table

Fig. 3.58 CSV complete

Check your understanding 15

1 Create a new file named *Food*.

2 Import the CSV file *Vegetables* and save it in a table named *Planting*.

3 Do not set a primary key.

4 Close the file.

Fig. 3.59 Planting

Reports

As well as printing out records in the form of a basic table or query, Access offers the option to create new objects known as **reports**. These can be based on either a table or a query and can display all or a selected range of fields. In Access 2007 they are created in the form of a table where you can select any column or row to format as you like.

As with other Access objects, you can use the wizard to create reports step by step, or use the automated facilities to create an **AutoReport**.

Certain entries may appear automatically – for example, totals for columns of numerical data – but these can be deleted just like any table entry, and you can also use the database tools to carry out your own calculations such as totalling or averaging a range of figures.

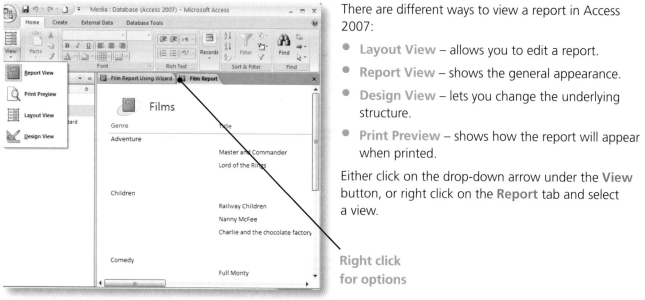

There are different ways to view a report in Access 2007:

- **Layout View** – allows you to edit a report.
- **Report View** – shows the general appearance.
- **Design View** – lets you change the underlying structure.
- **Print Preview** – shows how the report will appear when printed.

Either click on the drop-down arrow under the **View** button, or right click on the **Report** tab and select a view.

Right click for options

Fig. 3.60 View report

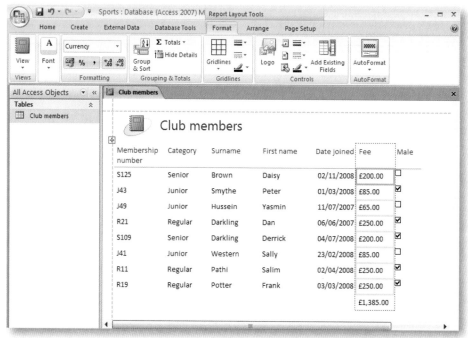

Fig. 3.61 A report

produce an AutoReport

1   Open the table or query on which to base the report.
2   On the **Create** tab, click on **Report**.

Report button

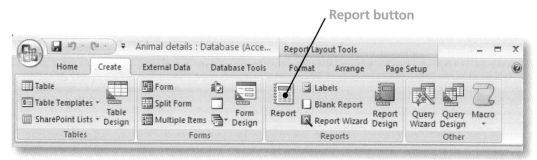

Fig. 3.62 Create tab with Report button

3   A new report will appear and the **Report Layout Tools** commands will be available.

format a report

1 Click on any field name or title to change the entry.

2 If the report is too wide, click on any heading or column of data to drag in the column boundary. You could also apply new formatting such as realigning headings or decreasing font sizes to reduce the size of the overall report.

3 Click on any numeric fields and use options from the **AutoSum** button to remove an unwanted total (by removing the tick next to its name) or to carry out a different calculation.

4 Click on any entry and press **Delete** or right click for the delete option to remove unwanted items such as a logo, header or footer text.

5 Select the entire table and add gridlines or borders by selecting one of the styles available.

6 Click on **AutoFormat** to select from a range of report styles.

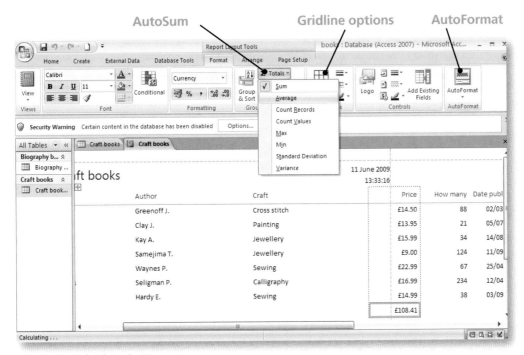

Fig. 3.63 Calculate in Report

7 When you try to close or if you click on the **Save** button, you will be asked to name the report. Type in a new name or it will have the same name as the table/query on which it is based. Its name will appear on a new tab showing the green Report icon.

To return to **Layout View**, right click on the tab in the navigation pane or select the option from the drop-down arrow below the **View** button.

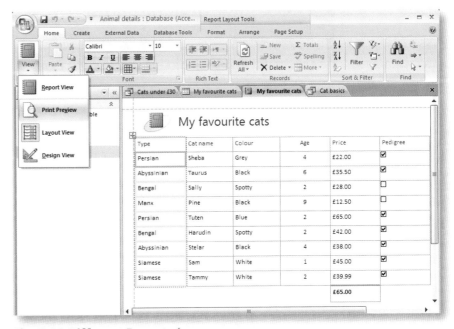

Fig. 3.64 Different Report views

Grouping and sorting

It may help make the report easier to understand if you group the records under particular headings. For example, you could group books by their category, villas by their country, team members by their sport or employees by their department. You can also reorder the records within each group by applying an ascending or descending sort.

add groupings and change sort order

1 Click on the **Group & Sort** command on the **Report Layout Tools – Format** tab.
2 Click on **Add a group** to display all the fields in the report.
3 Click on one to make it the first order category and the report will be organised under this grouping.
4 Click on **Add a sort** to select a field on which to sort the records.

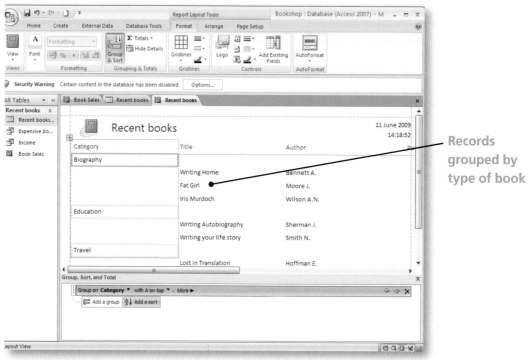

Fig. 3.65 Group & Sort

Check your understanding 16

1 Open the database file *Media* provided on the CD-ROM accompanying this book.

2 Create an **AutoReport** based on the *Films* table.

3 Include all the fields.

4 Group the report under *Genre*.

5 Sort the records in each group in descending order of *Bookings*.

6 Add a calculation to total all the *Bookings*.

7 Save as *Film Report*.

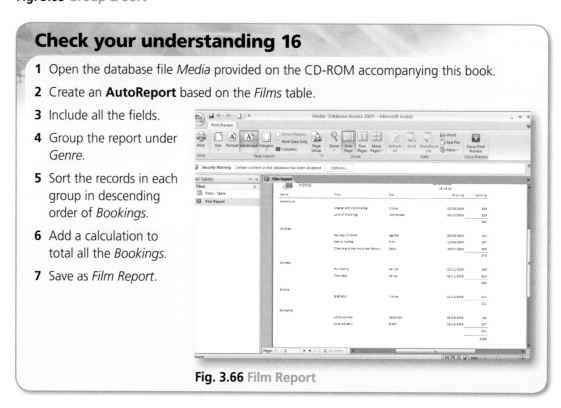

Fig. 3.66 Film Report

Using the wizard

When you want more control over the report, and to select your preferred field order or layout, you can use the wizard.

use the Report Wizard

1 Click on the **Report Wizard** command on the **Create** tab.
2 Select the table or query on which to base the report.
3 Add the fields you want to include by selecting them in the left-hand pane and clicking on the single arrow button.
4 To add all the fields in their current order, click on the double arrow button.
5 Work through the wizard by clicking on the **Next** button.

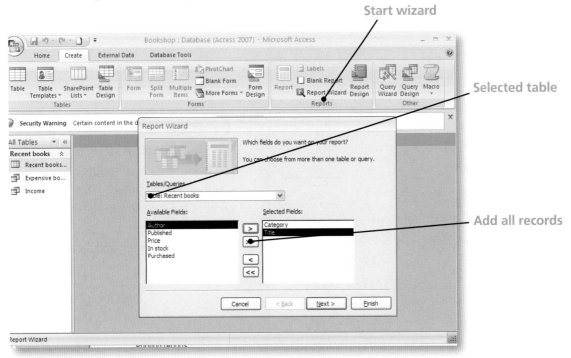

Fig. 3.67 Report wiz1

6 Click on a category in the left pane and then click on the single right-facing arrow if you want to group your records in the report.

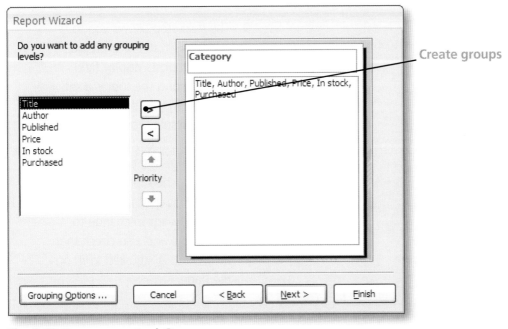

Fig. 3.68 Group report wiz2

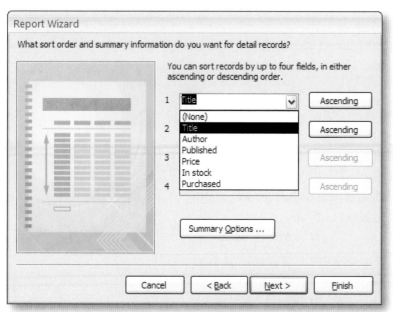

Fig. 3.69 Sort Report wiz3

7 To sort the records in a report, click in the drop-down box and select the first category on which to base the sort. Click for ascending or descending order. You can also add second and third order sorts if necessary.

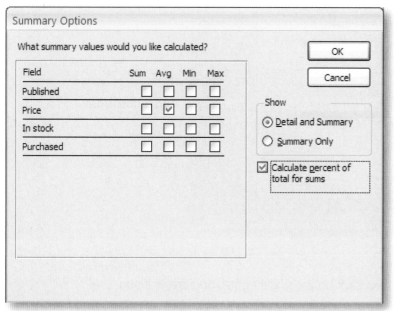

Fig. 3.70 Calculate Report wiz4

8 At this stage, you can click on **Summary Options** to add calculations to your report on suitable numeric fields. The choices are totals, averages, maximum or minimum and you can also show the results as a percentage.

Fig. 3.71 Report wiz5 layout

9 The next step allows you to select a layout. For example, tabular reports display field names along the top as a set of headings, whereas columnar reports display complete records one below the other. You can also change the orientation of the finished report. For a large number of field columns, you may want to change to Landscape and also check that field width is adjusted to fit on the page.

10   The final steps allow you to select a font size and style and name the report before you
click on **Finish**.

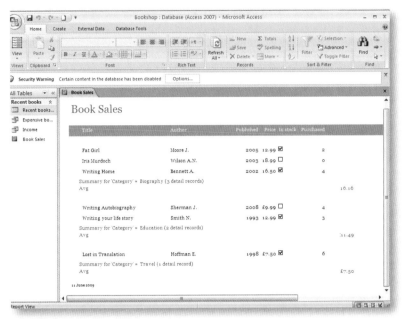

Fig. 3.72 Report wiz6

Creating labels

One option in the **Report** group on the **Create** tab is to produce labels based on a table or
query. Once again, work through the wizard to add and place data in the appropriate places.

1   Click on **Labels** on the **Create** tab.

2   Select the size of label you want.

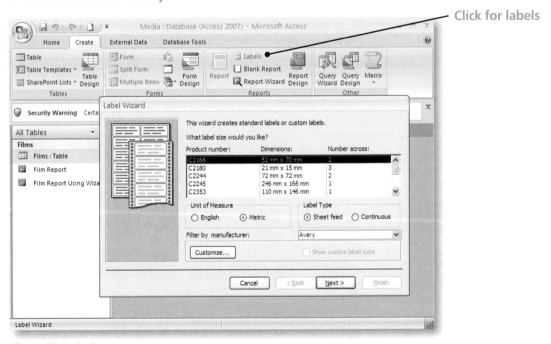

Click for labels

Fig. 3.73 Labels

3   Set the font style.

4   Use the arrow button to place fields in the right places on the label. Press **Enter** to move
down a line and use the spacebar to space entries correctly.

5   Select a field on which to sort the labels if necessary and then name the object.

6   Click on **Finish** to see the labels as they will be printed.

7   If they do not all fit on a single page when viewed in **Print Preview**, go to **Page Setup**
and limit the number or width of the columns.

Making amendments

When a report appears, particularly after working through the wizard, you may find things wrong with it. These can include:

- too much data relating to any summary options selected
- some columns too close to others
- the wrong wording for automatic entries.

To correct errors, you can reduce the width of columns, realign entries, amend text and change field properties. Changes can be made in either **Design** or **Layout view** and you may prefer to start changing things in **Layout view**. Where you want to make more detailed changes than are possible here, go to **Design view**.

In this view you will see that a report is made up of sections: the **Page header and footer** areas that will show entries on printed pages, and the **Report header and footer** areas that will show entries at the beginning and end of a report on screen and on the first or last page printed.

In the centre is the **Detail** section containing field names and the data drawn from the database.

make changes in Design view

1   Right click on the report tab and select **Design view** or select this view from the drop-down arrow below the **View** button.

2   If columns are too close, click to select any boxes (known as **controls**) containing field names and the data drawn from the database and drag them left or right.

3   You can also click on a control and drag the boundary in if it is too wide for the entry.

4   To reorder the fields, select any field and the related data and drag them to a new position. This will be marked by a vertical orange line.

5   To add extra entries such as a subtitle, click on the **Label** button **Aa** and draw a box before entering and formatting the text.

6   Click on and delete any unwanted entries such as the date or time, page number or a description of a calculation. Do *not* delete the actual formula or function – for example, the box containing an entry such as **=SUM**.

7   Click in any box and edit or format the text. For example, the label for an average or total will need to be made clearer as all it will say is **Avg** or **Sum**.

8   To change the title, there is a **Title** toolbar button you can click on to select it quickly.

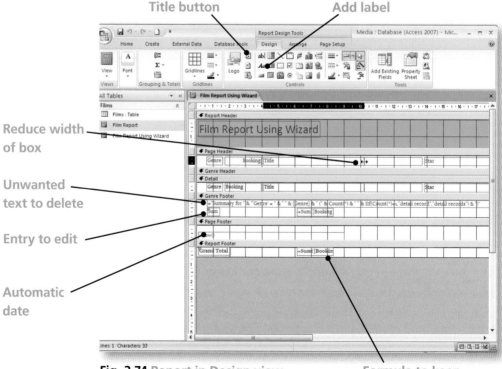

Fig. 3.74 Report in Design view

Headers and footers

As with other Microsoft Office documents, you can add extra entries to the top or bottom margins that will not change the layout of your main report. Note that some of these entries will have appeared automatically and you can choose to keep or delete them.

add headers or footers

1 Open the report in **Design view**.

2 On the **Report Design Tools – Design** tab, add an automatic entry by clicking on the **Insert Page Number or Date & Time** buttons.

3 The date and/or time will be added to the Report header area.

4 You can choose where to position the page numbers, and whether to include the number of pages. The entry will be added to the Page header or footer.

5 Add other entries such as the name of the report author by drawing a label and entering the text into the header or footer area of choice.

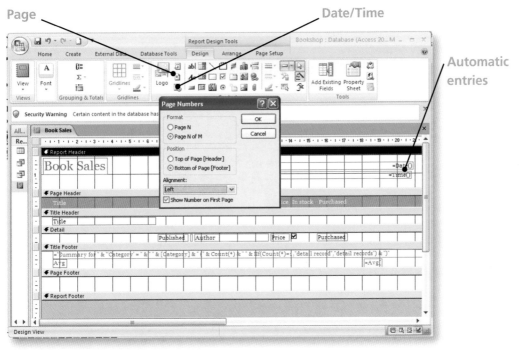

Fig. 3.75 Page Numbers

Check your understanding 17

1 Open the *Media* database.

2 Make the following changes to the *Film Report*:

a Move *Showing* so that it appears after *Booking*.

b Reduce the width of the *Star* and *Booking* columns.

c Change the title to *Film Details*.

3 Remove the time automatically added to the Report header.

4 Save and close the file.

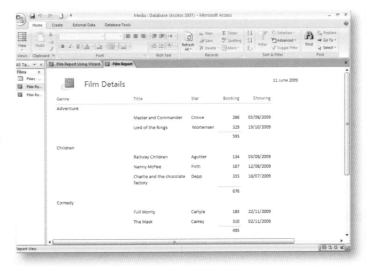

Fig. 3.76 Changed report

Adding the filename

Unlike in Word, there is no automatic entry for fields such as the filename in Access. To add this, you need to insert a control and type in a code to add the name of the field manually.

add the filename

1 In **Design view**, click on the **Text box** button **ab** and draw a box where you want the entry to appear. This normally allows you to add controls that will draw data from the underlying table.

2 It will contain the word 'Unbound' as it is not yet linked to any field in the database.

3 A second, smaller box will also appear labelled *Text* that is normally used as a label for the field. In this case, it can be deleted.

4 Right click on the **Unbound** box and select **Properties**.

5 A **Property Sheet** will open showing a list of all the fields in the database table. There is no entry for the filename so you need to add this yourself.

6 Click on the **Control Source** box and type in the correct code for adding the filename. This is in the form: **=CurrentProject.Name**.

7 Close the **Property Sheet**.

8 When you view your report in **Layout** or **Form view**, the filename will now be visible.

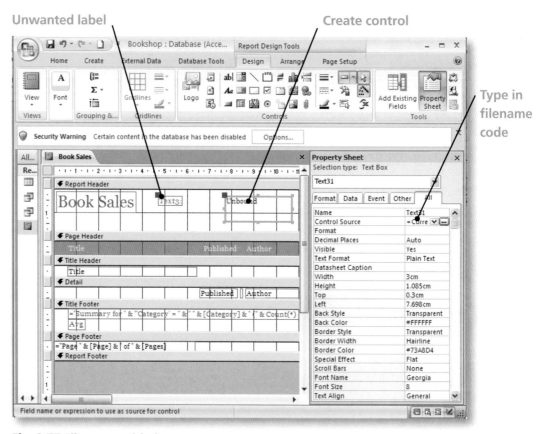

Fig. 3.77 Filename added

Formatting numeric data

You can also use the **Property Sheet** to make changes to the way numerical values are displayed. As during the design of a table, you can change the number of decimal places or add currency symbols and so on by changing the properties of the field.

format numeric data in a report

1 In **Layout view** click on the **Report Layout Tools – Format** tab and then click on an entry and use the formatting tools on the toolbar to set basic properties.

   or

2   In **Design view**, right click on the control containing the values drawn from the table.
3   Select **Properties**.
4   The name of the control will show at the top of the box.
5   In the **Format** or **Decimal Places** boxes, click on the arrow and select your preferred format.
6   Close the sheet or change view to see the effect.

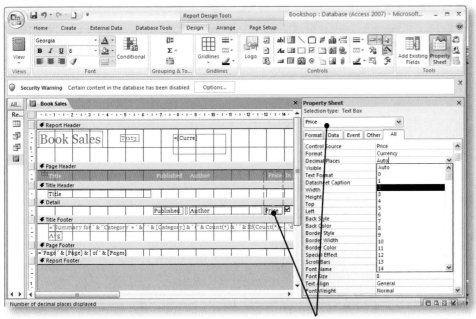

Fig. 3.78 Format numbers   Control being amended

Printing reports

Before printing, it is important to proofread and check for any errors. You should be familiar with the spellchecker as it works in the same way for all Microsoft Office 2007 applications. In Access, it is found in the **Records** group on the **Home** tab or can be opened by pressing the function key **F7**.

It is also important to check how the report will appear on the page so you need to view it in **Print Preview**.

preview a report

1   Click on the drop-down arrow under the **View** button and select **Print Preview**.
2   Check the number of pages, for example by clicking on the **Two pages** button. If the report is spread across too many pages, you can change to Landscape orientation.
3   Click on the **Paper size** or **Margins** button to make changes directly, or **Page Setup** button to open the dialog box.
4   Click on the **Print** button if you are happy to print.
5   Close the preview by clicking on the **Close** button.

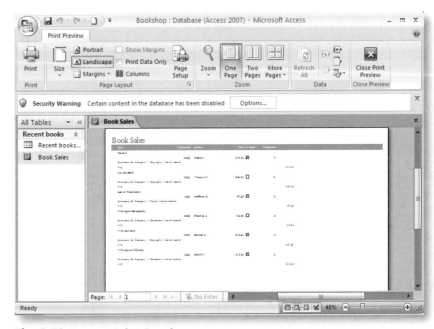

Fig. 3.79 Report Print Preview

Fitting reports to a specified number of pages

Sometimes reports spread onto another page despite changing orientation, or this may happen where you want to retain Portrait orientation. Here, you may need to go back into the design of the report to close up spaces before printing as required.

Whatever changes you make, always check that you have not lost any data that should be displayed in full.

reduce report size

1   In **Design view**, reduce the size of any control boxes where extra space is not required by dragging the boundary inwards.

2   Make sure the detail or footer areas have not been expanded too much – reduce them by dragging the boundary upwards.

3   Finally, move automatic entries such as page numbers that may have been placed too far to the right in a header or footer.

Printing

Once you are happy with the look of a report, print it in the normal way. The default settings will print in Portrait orientation.

print

1   Go to **Office – Button – Print – Quick Print** to print using the default settings.

2   Click on the **Print** option to open the **Print** dialog box.

3   Set the number of pages or copies you want and then click on **OK**.

Check your understanding 18

1   Open the file *Working lives* provided on the CD-ROM accompanying this book.

2   Create a report based on the table *Staffing*.

3   Save it as *Staff Report* and change the title to reflect this name.

4   Remove the *Surname* field.

5   Add a header containing your name.

6   Reformat *Salary* to show only whole numbers.

7   Remove the time which will have been added automatically to the header.

8   Display the average salary at the bottom of the *Salary* column and add a label *Average salary*.

9   Move *Region* so that it appears after *Age*.

10   Increase the height of the field name row so that *Length of service* appears on two lines.

11   Make sure you print out the report on a single page.

12   Print one copy of the report.

13   Close the file.

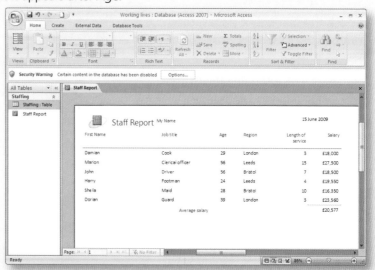

Fig. 3.80 Staff Report

CLAiT Assignment

CLAiT Plus Database Assignment

Task 1

1   Create a database file and name it **Entertainment**.

2   Create a table named **Theatre** containing the following fields:

FIELD NAME	DATATYPE
Theatre name	Text
Town	Text
Capacity	Number, 0 decimal places
Category	Text
Box price	Currency, 2 decimal places
Cheapest seats	Currency, 2 decimal places
Touring company	Yes/No (logic)
Date refurbishment starts	Date, English Long date format

3   Now enter the following records:

Theatre name	Town	Capacity	Category	Box price	Cheapest seats	Touring company	Date refurbishment starts
Tivoli	Leeds	550	Classical	£25.00	£6.50	☑	19 November 2009
Emporium	Bradford	885	Music hall	£32.00	£9.50	☐	02 January 2010
Regal	Burnley	325	Music hall		£4.99	☐	12 February 2010
Royal	Birmingham	885	Opera	£38.00	£9.00	☑	16 March 2010
Grand	Croydon	750	Classical	£22.50	£3.99	☐	08 March 2010
Coliseum	Manchester	996	Classical	£36.50	£9.95	☑	14 January 2010
Royal	Windermere	306	Opera	£25.00	£5.50	☐	28 November 2009
Winter gardens	Bournemouth	610	Light	£20.00	£7.50	☑	12 March 2010
Winter palace	Hastings	445	Light	£28.00	£10.00	☑	11 July 2010
Circle	Eastbourne	645	Classical		£8.75	☐	02 September 2010
Theatre upstairs	Stoke	250	Comedy		£3.75	☑	02 May 2010

Database records to enter

4   You need to use a code for the categories so change them as follows:

Classical – CL

Music hall – MH

Opera – OP

Light – LT

Comedy – CO

5   Use this table to create a tabular report named **Theatre details** in Landscape orientation displaying all the records.

6   Sort the records alphabetically by Town.

7   Include the date and a label saying **theatres** in the footer, and remove any page numbers.

8   Print a copy of the report on one page.

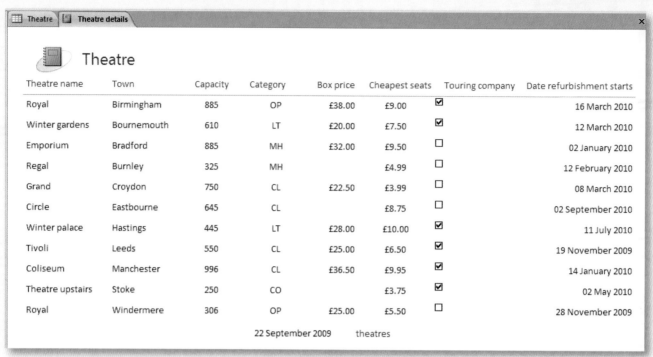

Theatre							
Theatre name	Town	Capacity	Category	Box price	Cheapest seats	Touring company	Date refurbishment starts
Royal	Birmingham	885	OP	£38.00	£9.00	☑	16 March 2010
Winter gardens	Bournemouth	610	LT	£20.00	£7.50	☑	12 March 2010
Emporium	Bradford	885	MH	£32.00	£9.50	☐	02 January 2010
Regal	Burnley	325	MH		£4.99	☐	12 February 2010
Grand	Croydon	750	CL	£22.50	£3.99	☐	08 March 2010
Circle	Eastbourne	645	CL		£8.75	☐	02 September 2010
Winter palace	Hastings	445	LT	£28.00	£10.00	☑	11 July 2010
Tivoli	Leeds	550	CL	£25.00	£6.50	☑	19 November 2009
Coliseum	Manchester	996	CL	£36.50	£9.95	☑	14 January 2010
Theatre upstairs	Stoke	250	CO		£3.75	☑	02 May 2010
Royal	Windermere	306	OP	£25.00	£5.50	☐	28 November 2009

22 September 2009   theatres

Report of theatres

Task 2

1 Create a set of labels containing the following fields set out as follows:

THEATRE NAME, TOWN

DATE REFURBISHMENT STARTS

2 Add your name and the date as a footer.

3 Sort the labels in ascending date order.

4 Print out a set of labels on one A4 page in Portrait orientation.

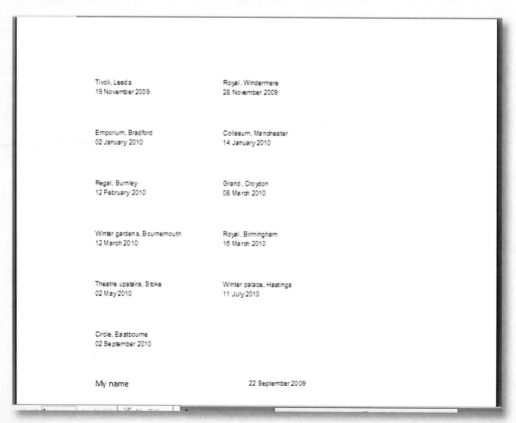

Labels

Task 3

1. Create a database file named **Tivoli Theatre**.

2. Import the file **Programme.csv** and name the table **Programme 2009**.

3. Check that the field names and datatypes are as follows:

Play	Text
Days of run	Number
Seats sold	Number
Average seat price	Currency, 2 decimal places
Full house	Yes/No (logic)
Start date	Date

4. Make the following amendments:

   a. Hamlet ran for 14 days

   b. The number of seats sold for Major Barbara was 4235.

5. Print out a copy of the table.

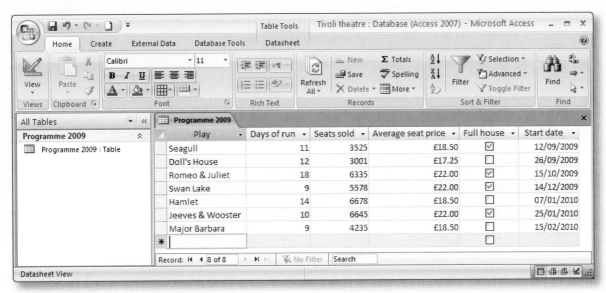

Imported table

Task 4

1. Create a query to identify all plays starting in 2010. Display only the **Play**, **Seats sold**, **Average seat price** and **Start date**.

2. Add a calculated field named **Average Income** (**Seats sold** x **Average seat price**).

3. Save this query as **2010 Income**.

4. Create a report based on this query that will display only **Play**, **Average Income** and **Start date**.

5. Group the report by **Start date** month.

6. Add subtotals and an overall total for the income.

7. Edit the labels as follows: subtotal label: **Monthly Income**, overall total label: **Final income**.

8. Amend the properties of these totals to show currency with no decimal places.

9. Remove unwanted text including the group heading "start date by month".

10. In the footer, display the date, remove the page numbers and add the text: **Bookings**. Position this label under the date.

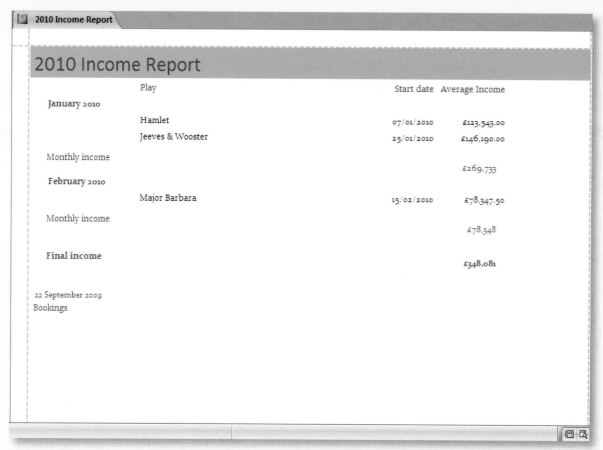

Income report

Task 5

1   Create a new query for all plays that ran for 10 or more days. Display the fields: *Play, Seats sold, Start date, Full house* and *Average seat price*.

2   Save as *Long run*.

3   Use this query to create a report displaying all the fields.

4   Group the records by *Full house*.

5   Sort first by *Average seat price* and then by *Play*, both in ascending order.

6   Save the report as *Long running plays*.

7   Amend field properties so that *Full house* shows as *Yes* or *No* and *Average seat price* displays two decimal places.

8   Adjust alignment so that column entries are clearly displayed.

9   Add your name to the footer and adjust the page number entry so that it does not show how many pages there are.

10 Print one copy.

11 Save and close all files.

Long running report

Designing an e-presentation

This unit involves the use of the presentation software Microsoft PowerPoint 2007 to create complex layouts and designs. You will learn how to use a master slide to maintain continuity throughout a presentation, how to amend and control slides and how to add interest when running a slide show.

At the end of the unit you will be able to:

⊕ use presentation software correctly

⊕ set up a master slide

⊕ create a presentation

⊕ insert and manipulate data

⊕ control a presentation

⊕ save, print and produce support documentation.

Launching PowerPoint

When the program opens, you will see one slide in the main window temporarily named Presentation1 plus panes offering different options. This is known as **Normal view**.

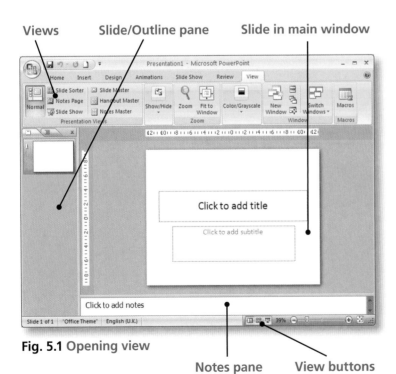

Fig. 5.1 Opening view

launch PowerPoint

1 Click on the **Start** button.

2 Go to **All Programs – Microsoft Office – Microsoft Office PowerPoint 2007**.

In **Normal view** you can:
- work directly on a slide
- enter text onto the slide in an **Outline** pane
- view thumbnails of any slide on a **Slide** pane (which alternates with **Outline** pane – click on the tab to change view)
- add speaker's reminders into the **Notes** pane.

On the **View** tab or from a button at the bottom of the slide you can also move to:
- **Slide sorter view** – an alternative way to view all the slides
- **Slide Show** – the view in which to run through a presentation on a computer
- **Slide Master** – this offers a way to create a template on which all the slides in the presentation will be based.

close a presentation

1 Click on the **Close** button.

or

2 Go to the **Office** button – **Close**.

3 To exit PowerPoint, go to the **Office** button – **Exit PowerPoint**.

Saving

As well as saving presentations as **.pptx** files, you can save them as shows, web pages, templates, images or other file types. Select the file type in the **Save as type:** box when saving.

save a presentation

1 Click on the **Save** button.

2 Select the location in the **Save in:** box.

3 Rename the file if necessary.

4 Change the file type if saving in a different format.

5 Click on **Save**.

Fig. 5.2 Saving

Slide layout

Slides come with pre-set areas for entering text or other content known as **placeholders**. The first slide is a **Title slide** and has areas for a title and subtitle. When you add new slides, these will have a **Title and Content layout** unless you have previously selected a different layout, in which case they will repeat the selected style.

change slide layout

1 Right click on any slide and select **Layout** or click on the **Slide Layout** button on the **Home** tab.

2 Scroll through the selection and click on your preferred layout.

3 This will now be applied.

Slide layout

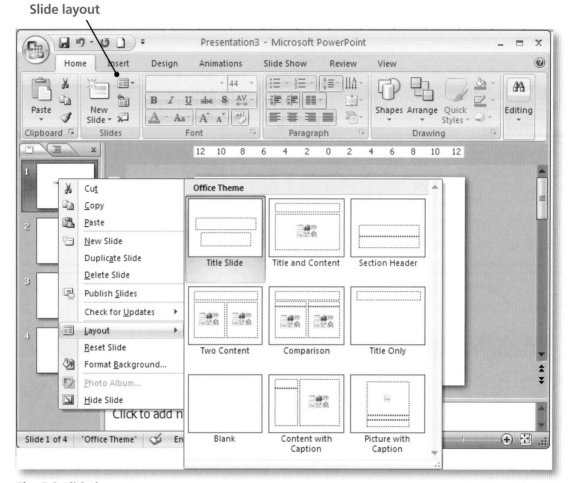

Fig. 5.3 Slide layout

Adding text

As well as typing in a text placeholder, you can add text to any part of a slide by creating a text box. In PowerPoint 2007 the default font is Calibri – size 44 for main headings, size 32 for subheadings and size 18 in a text box.

To make changes to these defaults, select the text and apply an alternative style, size, font colour or emphasis from the drop-down buttons on the **Home** tab. (See Unit 002 for details on text formatting.)

To set a size that is not displayed, type it over the size in the box and press **Enter** to confirm the setting.

add text to a placeholder

1 Click in the box and type your own entry.

2 To set text on a new line within the placeholder, press **Enter**.

3 If you type too much text for the box size, an **AutoFit Options** button will appear. Click on **AutoFit** to reduce the font size so that it fits the box, or click on **Stop** to leave the font size as set.

4 You can use the alignment buttons to set the text to the left or right of the placeholder.

5 Select any text and press the **Delete** key to remove unwanted entries.

Format entries

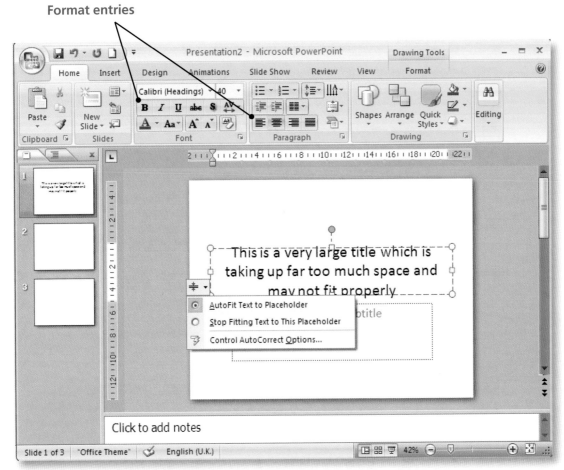

Fig. 5.4 Add text to a placeholder

add text in a text box

1 Click on the **Insert** tab.
2 Click on **Text Box**.
3 Move the pointer onto the slide and drag out a box.
4 The cursor will be positioned inside and you can start entering your text.

copy in text created elsewhere

1 Select the text in the original document.
2 Right click and select **Copy**.
3 Click on the '**Click on to add…**' text inside a placeholder.
   or
4 Click next to a slide icon on the **Outline** tab.
5 Right click and select **Paste**.
6 If there is a large block of text, you may need to select it and reduce the size to fit on the slide.
7 If you first right click on an empty part of a slide rather than inside a placeholder, the copied text will be pasted into a newly created text box. If necessary, click to select the border and resize it by dragging.

Pasted onto blank part of slide

Pasted into a placeholder

Fig. 5.5 Copied text file

Check your understanding 1

1 Open a new file and save it as *Conference*.

2 Open the Word document *Conference text* provided on the CD-ROM accompanying this book.

3 Select the title text and paste this into the **Title** placeholder in your presentation.

4 Now select the three selling points and paste these into the subtitle placeholder.

5 Save and close the file.

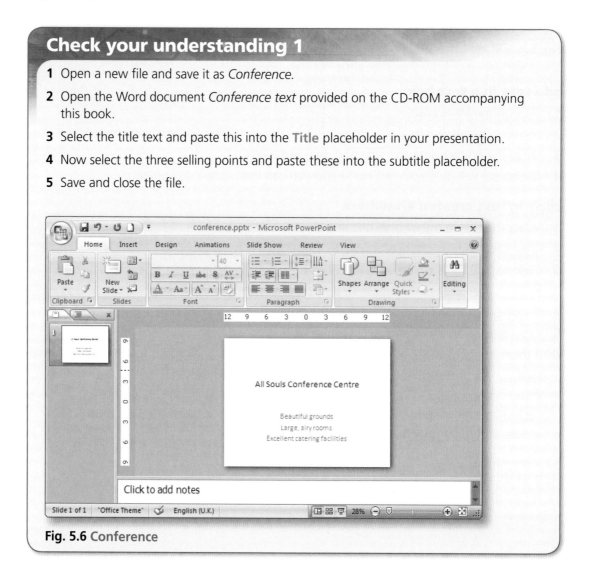

Fig. 5.6 Conference

Using Master slides

All Microsoft Office programs offer default settings that apply to various aspects such as font, font size, margins or orientation. In PowerPoint, the **Slide Master** stores information about the layout of each file. If you make changes here, they will be reproduced throughout the presentation.

Some of the common changes made to the master slide include colouring slide backgrounds or adding images or other objects (which are explained later in this unit). Follow exactly the same procedure but make sure you are in master slide view first if you want all the slides to have the same appearance.

You can also insert header or footer entries or apply different font styles to headings and subheadings. You will see that these are set at various levels, each one having its own indent and style of bullet point. To change any of these, first select the relevant level.

As you may want the title slide of a presentation to look different, there is a **Title Master** as well as **Slide Master**. To apply settings to every slide including the title slide, you must make sure you select the **Office Theme Slide Master** – the top slide in the pane – when setting up the master slide.

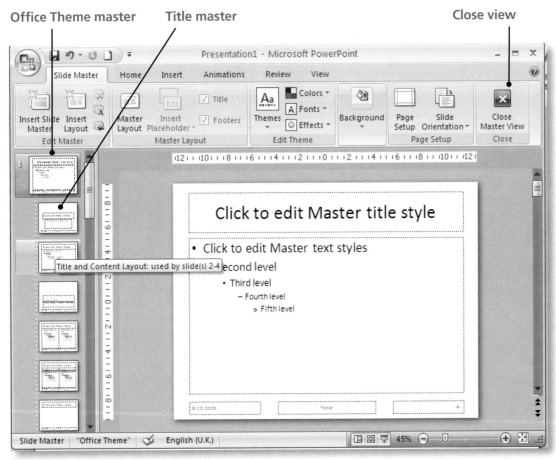

Fig. 5.7 Master

set up the master slide

1 Click on **Slide Master** on the **View** tab.

2 Click on the **Office Theme** thumbnail, or **Title Master** thumbnail to change that slide only.

3 Click on any text entry to apply different font styles, sizes, colours or bullet points from the **Home** tab in the normal way. Return to **Slide Master** view by clicking on the tab that will still be visible.

4 As an alternative, use the gallery to select a theme from the button on the toolbar. This applies an entire colour scheme that will combine backgrounds, fonts and borders and will influence the appearance of objects such as charts added later.

5 If required, change orientation or margins using options from **Page Setup**, available on either the **Design** or **Slide Master** tab.

6 Return to **Normal view** by clicking on the **Close Master view** button or the **View** button at the bottom of the screen.

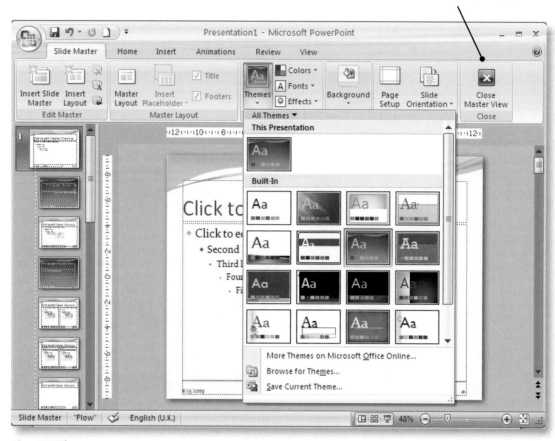

Fig. 5.8 Themes

Check your understanding 2

1 Start a new presentation.

2 In **Slide Master view**, select several different themes and see the effect.

3 Change the font type and size for the title text.

4 Change the slide orientation.

5 Return to **Normal view**.

6 Close the file without saving.

Using templates

If you want to use new designs more widely than in the current presentation, applying changes made to the master slide to future new slideshows, you can save the presentation as a template rather than a normal PowerPoint file. Your settings will now be available to use in any new presentation.

save a presentation as a template

1 Open the **Save As** dialog box.

2 Name the template.

3 Click on the **Save as type:** box and select **Template** as the file type.

4 A new templates folder will appear in the **Save in:** box.

5 Click on **Save** to complete the save.

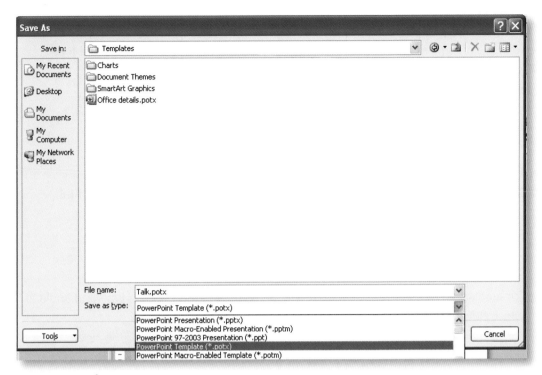

Fig. 5.9 Template

work with a template

1 Go to the **Office** button – **New**.

2 You can select a template from the gallery on which to base your new presentation, or locate a template you created and saved.

3 Click on the name of any template in the **Recently Used Templates** section, or click on the **My templates** link.

4 All the templates you have created will be displayed.

5 Click on one and then click on **OK**.

6 You can now make changes to any slides and save the presentation in the normal way.

View your own templates

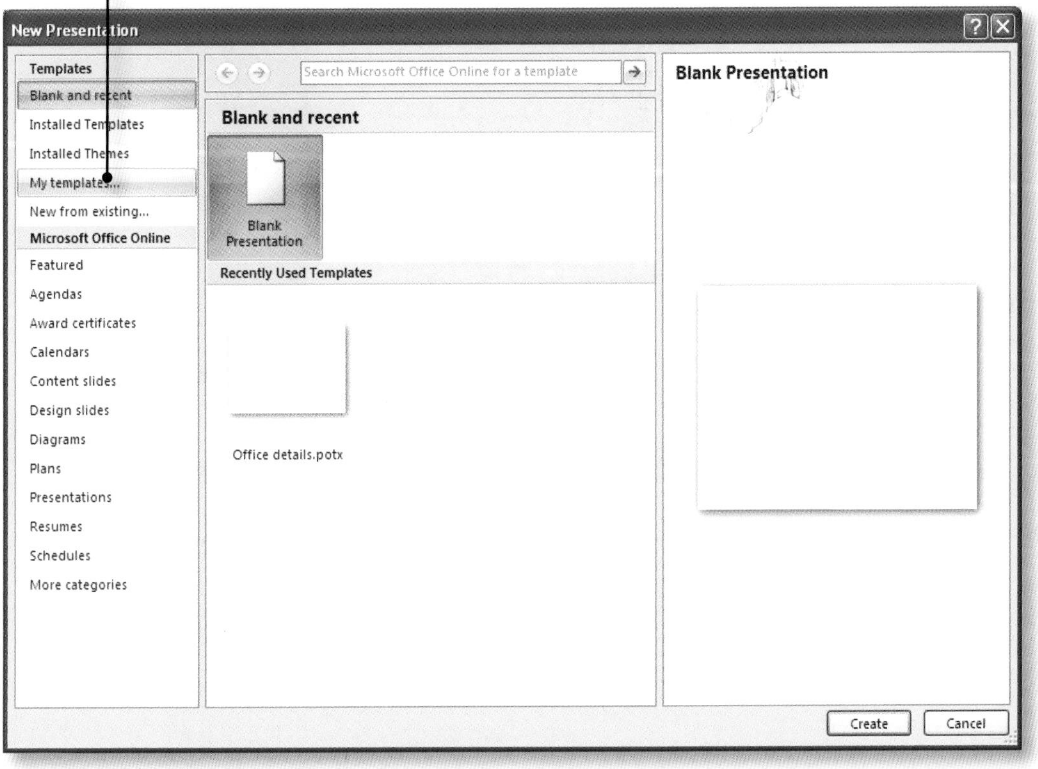

Fig. 5.10 Use template

Adding backgrounds

To colour one or more slides in a presentation, you can apply a background. There are four choices:

- solid colour
- a gradient of two or more colours
- a picture
- texture

apply a background

1 Click on the drop-down arrow in the **Background Styles** box on the **Design** tab.

2 Click on **Format Background**.

3 Select a type of background by clicking in the radio button.

4 For solid colours, click on the drop-down arrow in the **Colour** box to select from a limited number of colours or click on **More Colours** for a wider palette.

5 For gradients, choose a pre-set mix or select colours and then set the type and angle at which they merge together.

6 For pictures or textures, browse for a picture file or choose from the drop-down list of textures. You can also click on the **Clip Art** button to search for a **Clip Art** image.

7 If required, slide the transparency slider for a less opaque colour/picture.

8 To apply the background to a single slide, click on **Close**.

9 To apply the same background to the whole presentation, select **Apply to All**.

10 To return to the original background, click on **Reset**.

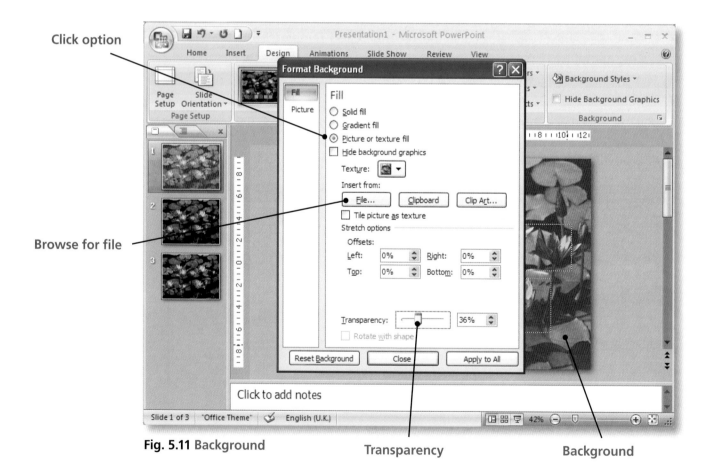

Fig. 5.11 Background

In the CLAiTPlus assessment, there is a tolerance of 10mm for a background image to fill the slide.

Check your understanding 3

1 Start a new presentation.

2 Save as *Clocks*.

3 Apply the image *clock picture* provided on the CD-ROM accompanying this book, as a background.

4 Set 50% transparency.

5 Save and close the file.

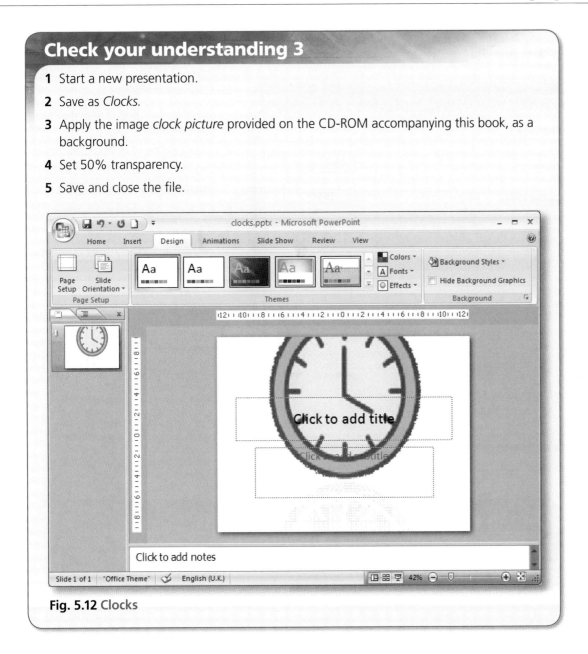

Fig. 5.12 Clocks

Headers and footers

If you want to add automatic entries such as the date or slide number, or type in your own text, you can do this by inserting footers in the bottom margin of the slides. You can also add header and footer entries to supporting documentation such as handouts or notes pages. The areas for different entries are pre-set on the slide, but you can drag the boxes to different positions and select and format entries using normal formatting tools.

add headers or footers

1 Click on **Header & Footer** on the **Insert** tab.

or

2 Click on the **Slide number** or **Date and time** commands.

3 When the **Header & Footer** box opens, click in the radio button for a fixed date and enter it in your preferred style.

4 For a date and/or time that will update automatically, click on the **update** option and select the style of entry you want from the drop-down arrow in the box.

5 Make sure the date is set as English (UK) or you will find the month precedes the day.

Check for time or
date formats

Change to UK

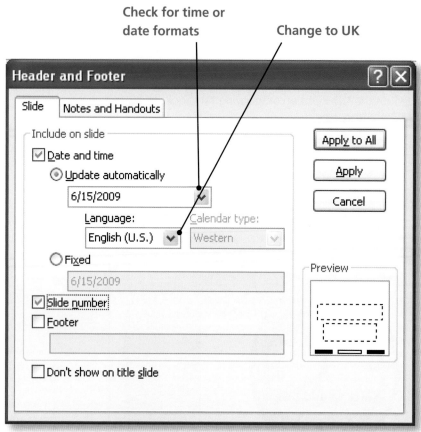

6 Click on the **Slide number** checkbox to add numbers to your slides.

7 Click on the checkbox and then type in any other entries in the **Footer** box.

8 Click on **Apply** for a single slide.

9 Click on **Apply to All** if you want the same entries on every slide.

Fig. 5.13 Header/Footer box

Check your understanding 4

1 Reopen the file *Clocks*.

2 Add the fixed date 12/2/2010 and a footer containing your name.

3 Select the entries and apply a bold emphasis.

4 Save these changes and close the file.

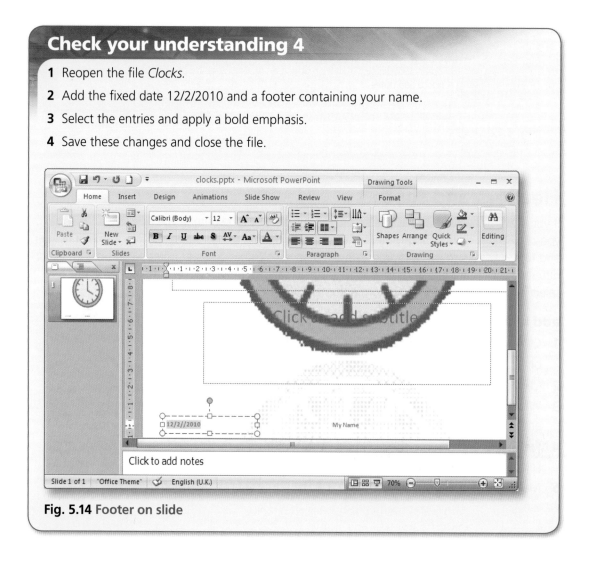

Fig. 5.14 Footer on slide

Text box properties

After adding text to a slide you can either display just the text or emphasise the text box/placeholder border or fill to make the text stand out more.

change text box properties

1 Click on the text to display the border and then right click on the line or inside the box.
2 The floating toolbar will appear and you can click on the drop-down arrow on the **Shape Outline** or **Shape Fill** button.
3 Choose a colour for the background or line and/or click on the arrow to open the gallery of line weights or styles.
4 Click on **More Lines** or **Colours** for further options.
5 Click on **No Outline or Colour** to remove the formatting.

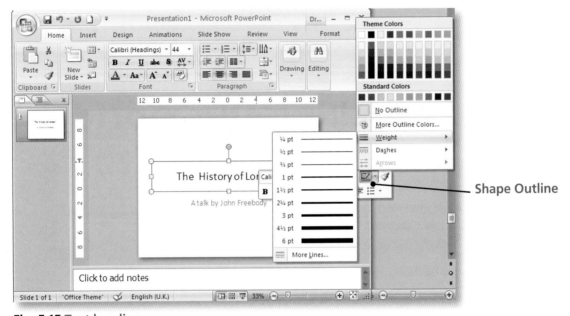

Fig. 5.15 Text box line

or

6 Find the **Shape Outline** and **Fill** buttons on the **Format** tab.
7 If you right click and select **Format Shape**, you will open the **Format Shape** dialog box.
8 Click on **Line** or **Fill** and select colours and styles from the various boxes.

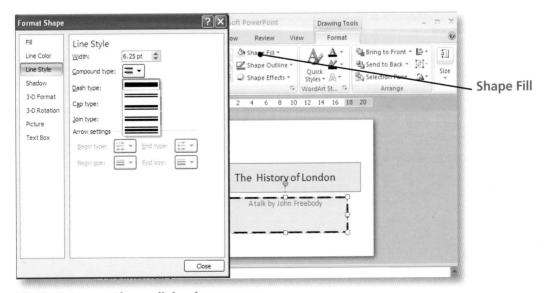

Fig. 5.16 Format Shape dialog box

Check your understanding 5

1 Start a new presentation and save it as *Travel*.

2 Add the title *Travelling the World* to the first slide.

3 Apply a different font type and set the size to 50.

4 Shade the placeholder pale green.

5 Add a thick or double line border in red.

6 Save the file.

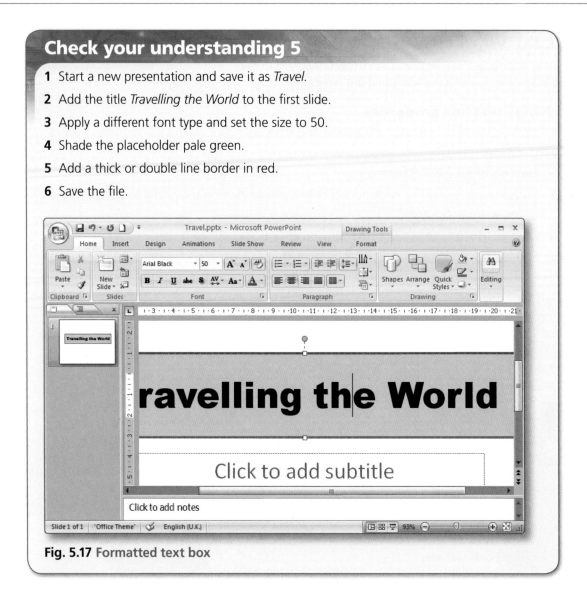

Fig. 5.17 Formatted text box

Spellchecker

Once you have typed or copied any text onto a slide, you need to check that it is correct. Proofread by eye for words that are spelt correctly but don't make sense, and use the spell-checker to correct any misspelt words.

spellcheck a presentation

1 For a single error, right click on any red underlined word and select from alternative spellings that may be offered.

2 To check the whole presentation, click on the **Spelling** button on the **Review** tab.

3 In the **Spelling** window, click on **Ignore** or **Ignore All** to retain a spelling that is correct.

4 Click on a suggested alternative or manually change a misspelt word in the **Change to:** box and click on **Change** or **Change All** to update the slides.

5 For words you want recognised in future, click on **Add** to add them to the dictionary.

6 Click on **Close** to return to your presentation.

Fig. 5.18 Spellcheck

Bullet points

As you type into some placeholders, bullets will be added automatically. You can remove unwanted bullets or add them where needed. Each level of text has its own bullet style applied automatically. Numbering works in exactly the same way as bullets so you can number list items instead.

work with bullets

1 Select one line, or all list items.

2 Click on the **Bullets** button on the **Home** tab to add bullets.

3 Click on the highlighted **Bullets** button to remove bullets.

4 Click on the drop-down arrow next to the button to select alternative styles of bullet.

5 If you change text level within a list, each level will have a different style of bullet applied by default. It will also be indented.

- To go down a level, press the **Tab** key.
- To go up, hold **Shift** as you press **Tab**.

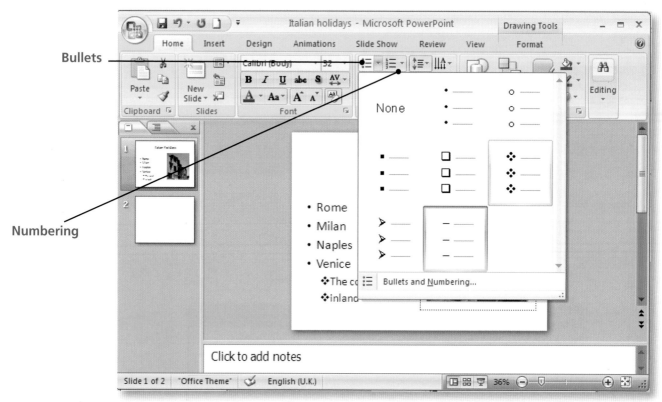

Fig. 5.19 Add bullets

6 Sometimes the bullet points or list items will be in the wrong place on the slide. For example, the text may be too close to the bullet. Change the position by dragging the bullet point or text insertion point along the ruler with the mouse. A dotted line will show the new position.

7 To increase spaces between entries in a list, apply different line spacing.

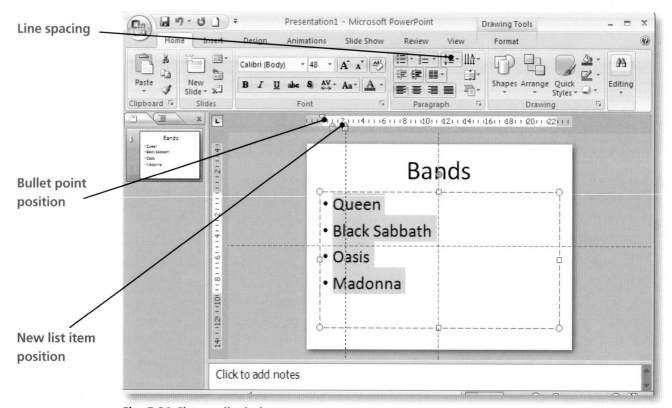

Fig. 5.20 Change list indent

add extra items to a bulleted list

1 Click at the end of an item and press **Enter**.

2 Type in the new entry.

3 To create an entry at a lower level, first press **Enter** to move to the line and then press the **Tab** key.

4 Select individual level entries to apply a different bullet style.

Check your understanding 6

1 Reopen *Travel*.

2 Add the following bulleted list to the slide, left aligned, in the subtitle placeholder:
- UK
- Europe
- Africa
- India
- The Far East

3 If necessary, drag the title placeholder higher up the slide to make room.

4 Change the style of bullets that have appeared by default.

5 Apply Courier, font size 36 to the list text.

6 Now add two items with a different style of bullet, indented at a lower level under Africa:
- Egypt
- South Africa

7 Save the file as *Travel bullets*.

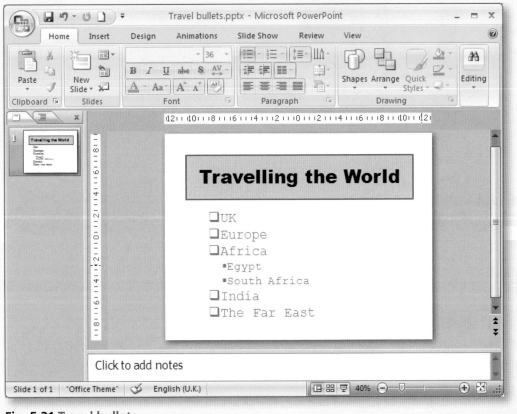

Fig. 5.21 Travel bullets

Building up a presentation

Most presentations are made up of a number of slides and these can be reordered, duplicated or deleted as you work on their contents.

add new slides

1 In the **Slide** pane, click on a thumbnail and press **Enter**.

or

2 Click on the **New Slide** button.

3 Choose a layout and the new slide will appear as the one *after* the slide on screen.

4 Move between slides by clicking on the number in the **Slide** pane or using the **Previous** and **Next** navigation buttons on the right-hand side of the slide.

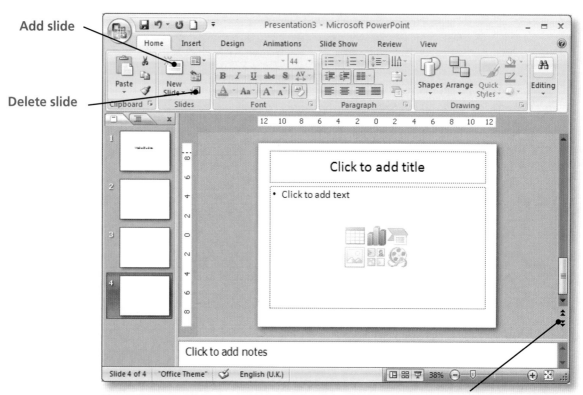

Fig. 5.22 New slide

delete a slide

1 Click on the slide in the **Slide** pane.

2 Press the **Delete** key.

or

3 Open the slide on screen and then click on the **Delete Slide** button showing a red cross.

Duplicating slides

You may want to create slides that are very similar. Rather than designing each one from scratch, you can duplicate them and then make the minor changes necessary.

duplicate a slide

1 Select the slide(s) you want to duplicate in the **Slide** pane.

2 On the **Home** tab, click on the drop-down arrow below the **New Slide** command.

3 Select **Duplicate Selected Slides**.

4 Copies of the slide(s) will appear as the next slide(s) in the presentation.

5 To copy a slide from one presentation to another, use **Copy and Paste**.

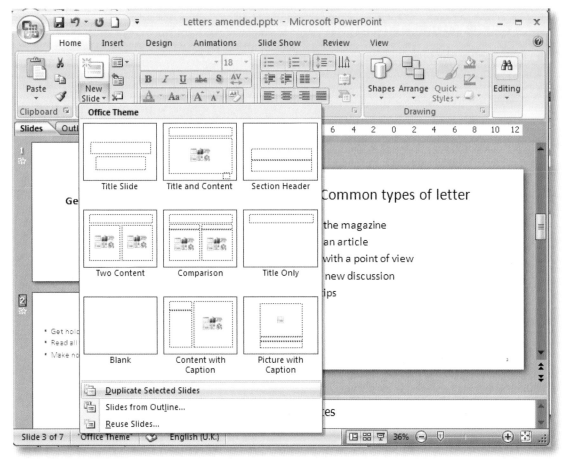

Fig. 5.23 Duplicate

Reordering slides

Having added a number of slides and decided on their content, you may want to display them in a different order. You can change slide order in two views: on the **Slide** pane in **Normal view** or in **Slide Sorter view**.

New position for slide

Slide Sorter view

Fig. 5.24 Reorder slides

reorder slides

1 Click on the slide you want to move on the **Slide** pane or in **Slide Sorter view**.

2 Hold down the mouse button.

3 Drag the slide to its new position.

4 This will be marked by a red vertical or black horizontal line.

5 Let go of the mouse and the slide will drop into place.

Check your understanding 7

1 Reopen *Travel bullets*.

2 Add two new slides.

3 Add the following text to slide 2:

What to pack (title)

- Clothes
- Documentation
- Washing things
- Things to do

4 Add the following text to slide 3:

Where to go (title)

- **Museums and galleries**
- **Train or boat trips**
- **The sights**
- **Towns and villages**

5 Now reorder the slides so that slide 3 becomes slide 2.

6 Duplicate slide 1 and move this slide to become the last slide in the presentation.

7 Now delete the duplicated slide.

8 Save the file as *Travelling* and close.

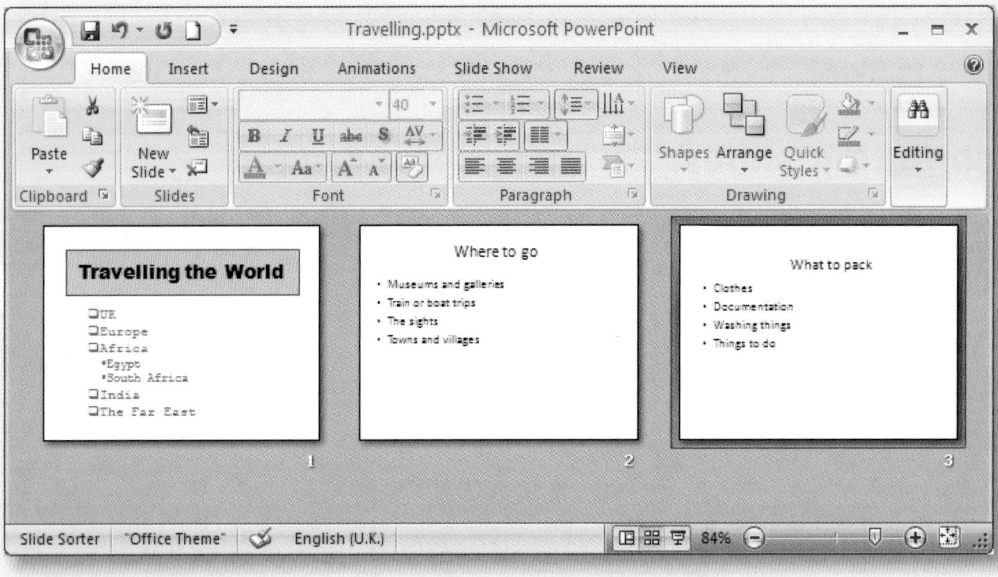

Fig. 5.25 Travelling

Find and replace

As with all Microsoft Office programs, there are useful search tools available to help you locate or make changes to text in your presentation. In PowerPoint they work in exactly the same way as in Word but there are less options to choose from.

find data

1 Click on the **Find** command on the **Home** tab in the **Editing** section.

2 Enter the text you want to find in the **Find what:** box.

3 You can click on the checkboxes to match case exactly and/or to find whole rather than parts of entries.

4 Click on **Find Next** to locate the first matching entry and keep clicking on the button to work through the presentation.

replace entries

1 Click on the **Replace** command on the ribbon or click on the button in the **Find** dialog box.

2 Type in the entry to be replaced in the **Find what:** box.

3 Enter the replacement text in the **Replace with:** box.

4 Set any options such as matching case.

5 Click on **Replace All** if you are sure you have entered the items correctly.

6 Click on **Find next** to locate the first matching entry.

7 Click on **Replace** to replace it or **Find Next** to leave it in the presentation and move on to the next matching entry.

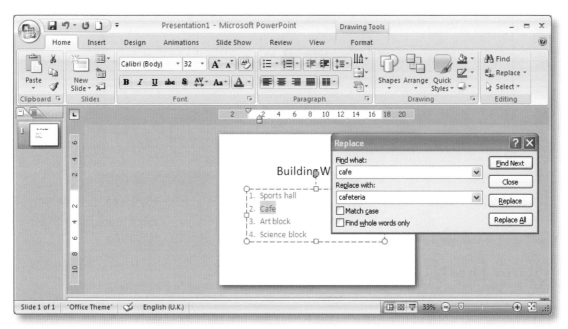

Fig. 5.26 Replace

Adding objects to slides

There are a number of additions you can make to your presentation to bring it to life or make the information it contains clearer or more memorable. These include:

- graphical images
- WordArt (which enables text to be shaped and textured)
- drawn shapes
- charts
- tables
- diagrams
- sounds
- animated images (labelled 'movies' in Microsoft programs).

Note that sounds and animations will only work in **Slide Show** view.

Depending on the type of object, once it is on a slide it may be possible to resize, rotate, align, layer, copy, move or group it with other objects, or format it by adding coloured borders or fills.

Content placeholders

When you apply a slide layout containing a **Content** placeholder, you will see small icons representing many of these objects. Click on any of the icons to start inserting that particular type of object. You will be able to either create the object directly or search the computer for files of that type.

Create a table   Create a chart   Create a SmartArt graphic, e.g. organisation chart

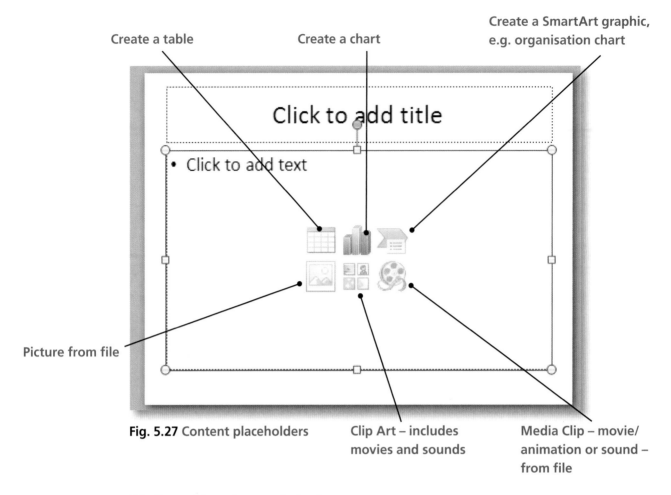

Picture from file

Fig. 5.27 Content placeholders

Clip Art – includes movies and sounds

Media Clip – movie/ animation or sound – from file

Using the Insert tab

If you click on the **Insert** tab you will see many of the objects listed. Click on the command to open an appropriate dialog box or menu.

Object to insert

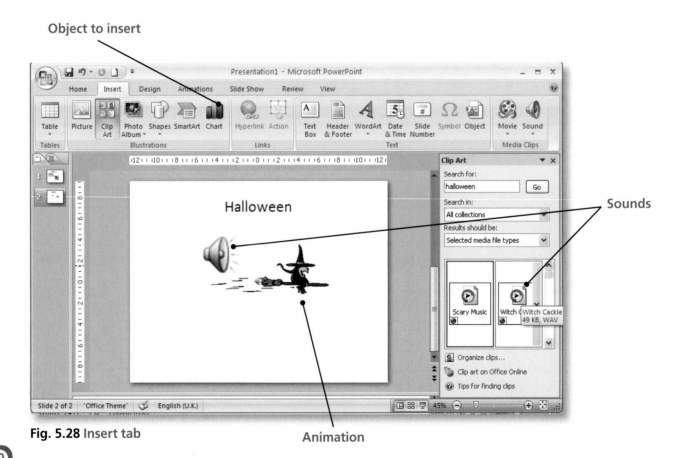

Sounds

Animation

Fig. 5.28 Insert tab

insert a Clip Art image

1 Click on the **Clip Art** command on the **Insert** tab.

2 Click on the **icon** in a **Content** placeholder.

3 A search pane will appear.

4 Enter your keywords and click on **Go** to locate appropriate images.

5 Scroll through the images.

6 Click on an image to add it to your slide.

insert a picture from a file

1 Click on the **Insert Picture** file command.

or

2 Click on the **Content** placeholder icon.

3 Browse through your folders to locate the file.

4 Click on its name and then click on **Insert**.

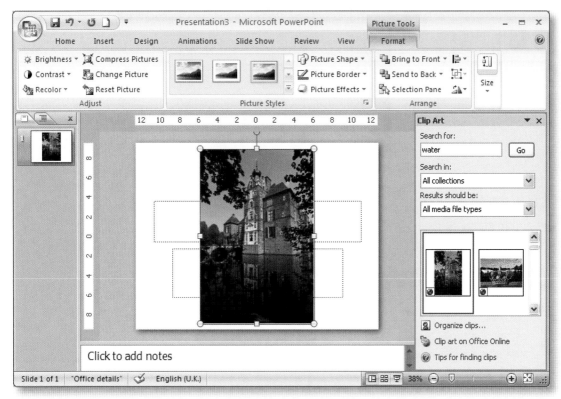

Fig. 5.29 Insert picture

Copy and Paste

You will often find it quicker to copy and paste an image. For example, any copyright-free images from the Web can be saved onto your computer or storage media to be inserted from file, or you can copy and paste them directly.

Formatting pictures

When a picture appears on the slide, it is in a selected state and shows small, white sizing handles round the edge plus a green rotate circle on an arm.

resize a picture

1 Make sure it is selected.

2 Position the pointer over a sizing handle. Select a corner position to keep the picture in proportion.

3 Gently click and hold down the mouse button as you drag the border in or out.

4 Let go when the image is the correct size.

or

5 On the **Picture Tools – Format** tab, enter exact measurements in the height or width boxes in the **Size** group.

or

6 Right click on the picture and select **Size and Position** to open the dialog box.

7 Click on the **Size** tab and change height or width measurements.

8 You should see that there is a tick in the **Lock aspect ratio** box which will maintain the picture's proportions.

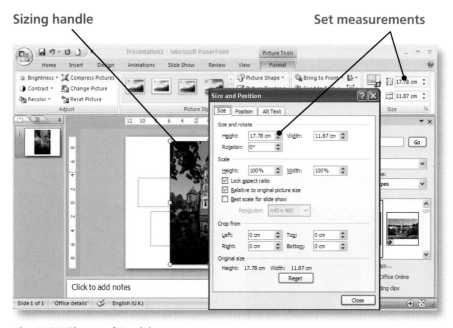

Fig. 5.30 Size and Position

Fig. 5.31 Position

position a picture exactly

1 Open the **Size and Position** dialog box.

2 Click on the **Position** tab.

3 The object can be positioned with a set measurement from the top left-hand corner (or centre) of the slide horizontally and/or vertically.

move a picture

1 Gently drag the picture to a different position when the pointer shows a four-headed arrow.

or

2 Right click and select **Cut**.

3 Click on the same slide or select a different slide to receive the picture.

4 Right click and select **Paste**.

copy a picture

1 Right click and select **Copy**.
2 Click on the slide to take off the selection, or click on a new slide.
3 Right click and select **Paste**.
4 You can also drag the picture as you hold down the **Ctrl** key to create a copy on the same slide.

delete a picture

1 Select the picture and press the **Delete** key.

Cropping

If a picture includes an unwanted portion, this can be removed by a method known as **cropping**. (It is not the same as resizing as that leaves all parts of the picture still visible.)

crop a picture

1 Select the picture.
2 Click on **Crop** on the **Format** tab.
3 This places black lines round the picture and the pointer will show the cropping shape.
4 Move the pointer to the border nearest the unwanted part and click on and drag the black shape.
5 Slowly drag this inwards or upwards.
6 Let go when the unwanted part of the picture disappears.
7 You can now work with what is left.
8 If you make a mistake, click on the **Crop** command again and drag the boundary back to its original position.

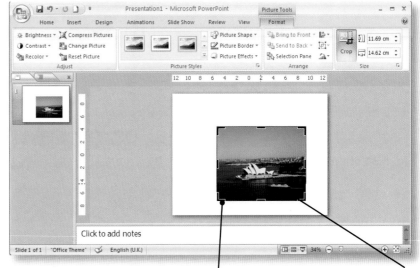

Fig. 5.32 Crop1   Black shapes to drag   Unwanted small boat

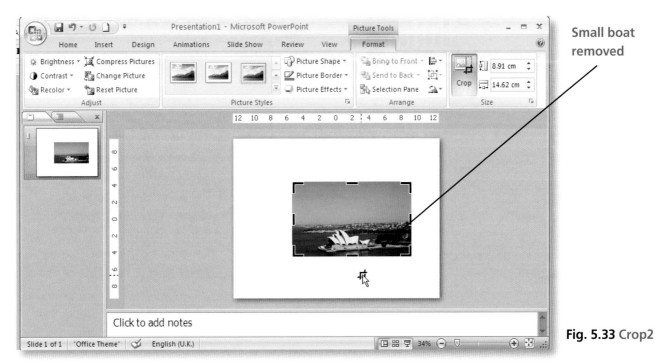

Small boat removed

Fig. 5.33 Crop2

Drawn shapes

PowerPoint 2007 contains a range of shapes and lines that you can add to a slide very quickly. You can then resize, rotate, format or group the shapes to build up complex and attractive drawings.

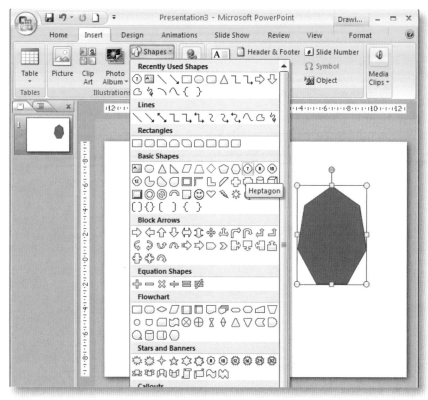

Fig. 5.34 Shape

add a line or shape

1 On the **Home** tab in the **Drawing** group, or from the **Insert** tab, click on the **Shapes** command.

2 Click on any style of line or shape to select it.

3 Click on the slide to add a shape or line at the default size.

4 Move the pointer to the slide and click and drag out the shape to set your own size.

5 When you let go, the shape will appear selected with a blue fill by default.

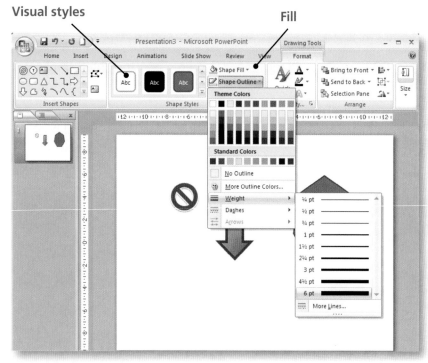

Fig. 5.35 Shape Outline

format a shape

1 Select the shape.

2 Click on the **Format** tab or use commands on the **Home** tab in the **Drawing** group.

3 Rest the pointer on one of the visual styles in the gallery to preview the effect and click on any to set a pre-defined design to your shape.

4 If you prefer, choose a different colour fill or line style and weight from the drop-down **Shape Fill** and **Outline** boxes.

5 For an arrow (not a block arrow), click on **Arrows – More Arrows** from the **Shape Outline** box to change the beginning or end type or size of arrowhead.

6 Click on **Effects** to apply shadows or reflections.

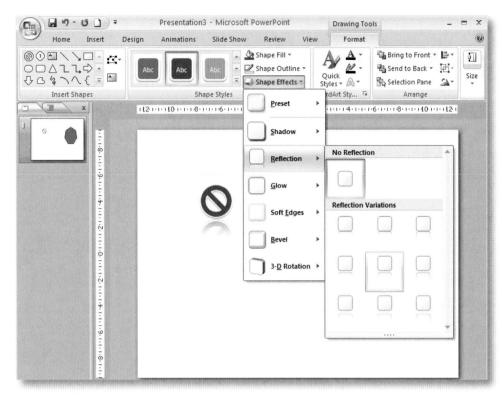

Fig. 5.36 Format shape

Adding text to shapes

If you want a message inside a shape, it is easy to add text.

add text to a shape

1 Click on the shape and enter your text.
2 On the **Home** tab, click on the **Text direction** box to select **vertical** or **stacked** text.
3 Click on **More Options** to set the text direction.
4 To edit the text, click on it and delete or amend as normal.
5 You can also select it and change alignment, font, font size or colour from commands on the **Home** tab. (A **Text Alignment** button is below **Text Direction**.)

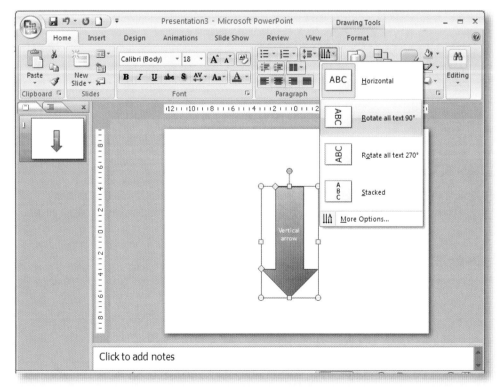

Fig. 5.37 Text in shape

Check your understanding 8

1   Start a new presentation.

2   Add a circle, square and triangle of any size.

3   Make the circle quite small and colour it red.

4   Colour the square yellow.

5   Make the triangle very large, colour it green and border it with a thick, black solid line.

6   Join the shapes with single arrows.

7   Increase the line weight for the arrows and colour them pink.

8   Add the text *Simple shapes* inside the triangle and position it so that it reads vertically upwards.

9   Increase the text font to size 36.

10   Close the file without saving.

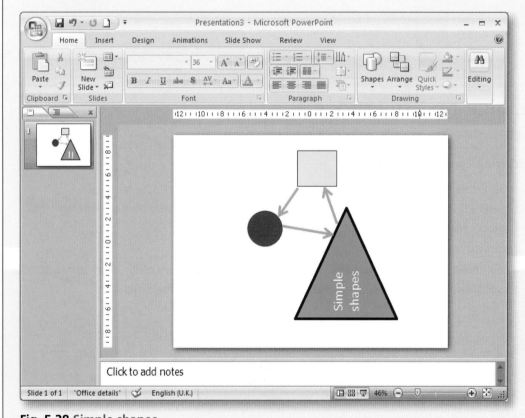

Fig. 5.38 Simple shapes

Rotating shapes

A selected shape shows a rotate arm that ends with a green circle. You can use this to drag the shape round manually, or set an angle of rotation. You can also flip shapes to create mirror images.

rotate a shape

1   Select the shape and position the pointer over the green rotate circle. You will see a black circle.

2   Hold down the mouse button and drag the shape round to a new position. The pointer will now display four black arrows.

or

3 Click on the **Rotate** command on the **Format** tab.

4 Select an option such as **Flip Vertical** to invert it, **Flip Horizontal** for a mirror image or rotate the shape in one direction by 90 degrees.

5 Click on **More Options** to open the dialog box where you can enter an exact angle of rotation.

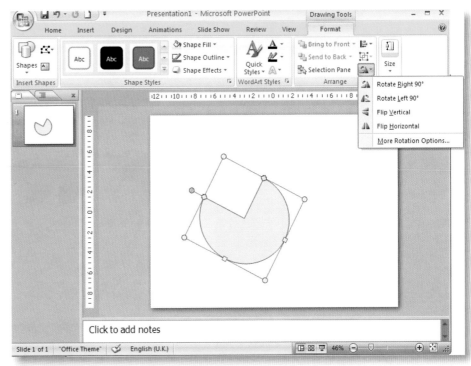

Fig. 5.39 Rotate

Multiple shapes

Once you have added a number of shapes to a slide, you can work with them in several ways:

- Align them to each other or the edges of the slide.
- Group them into a single drawing.
- Layer them on top of one another.

align shapes

1 Select one or more shapes.

2 Click on the **Align** command on the **Format** tab.

3 Select an alignment option.

4 For example, **Align Top** will move the selected shapes up to the top of the slide.

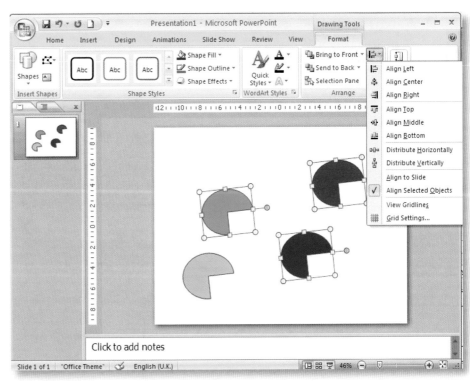

Fig. 5.40 Align1

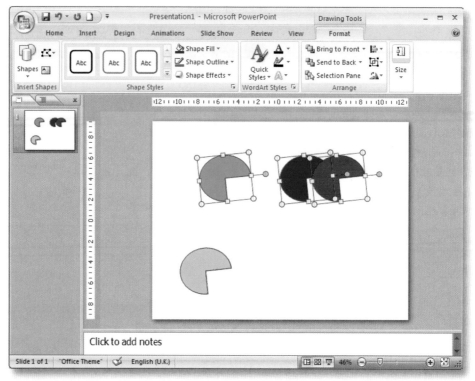

Fig. 5.41 Align2

Drawing guide   Grid

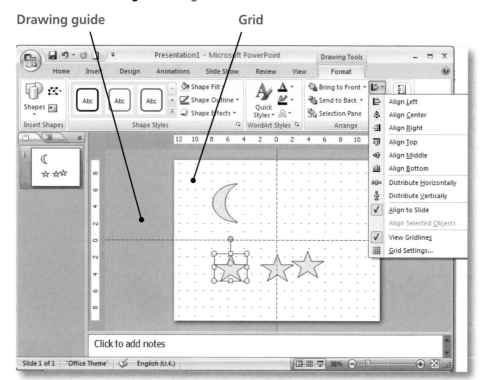

Fig. 5.42 Use grid

5 To help position objects by eye, you can display four drawing guides across the slide, or a set of gridlines. You can then drag objects and line them up with the lines and rulers.

6 If no rulers are visible, click on the checkbox on the **View** tab.

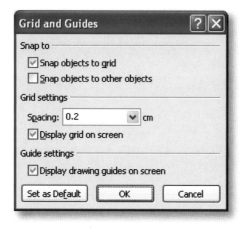

7 To open the **Grid and Guides** dialog box, right click on the slide or click on **Grid settings** from the **Alignment** command.

8 Remove the ticks in the boxes if you want to remove grids and guidelines.

Fig. 5.43 Grid and Guides

group shapes

1 Select the shapes you want to group. Either click on the first and then hold down **Ctrl** as you click on the others or draw round them with the pointer.

2 You can also include other objects such as WordArt or text boxes and group them with shapes.

3 Click on the **Group** option from the **Group** command on the **Format** tab.

4 You will see that all the shapes have been grouped into a single selected box and can be treated as one to copy, move or resize.

5 You will still be able to click on and select an individual shape to reformat it, even after it has been grouped.

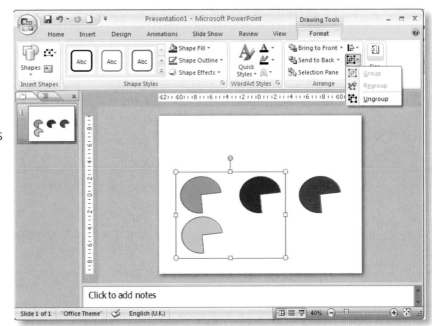

Fig. 5.44 Group shapes

6 As an alternative, you can click on **Ungroup** to make changes to parts of the grouped drawing, and regroup them again by clicking on **Regroup**.

Layering

An alternative way to build up a complex drawing is to layer shapes on top of one another. Once they are in place, an individual shape can be repositioned – either sent behind a single shape (backwards) or to the back of the stack, or brought forward in the same way. If you set a transparency level, you will be able to see shapes below others that have been layered. This is a useful way to use coloured shapes to frame text or other objects.

layer shapes

1 Drag the shapes on top of one another.

2 Make sure you can select a small part of any shape.

3 Select one shape and reposition it by sending it behind or in front of one or all the other shapes. Either click on **Bring to Front** or **Send to Back** and select **to the back/front** or **backwards/ forwards**.

or

4 Click on the **Selection Pane** command, click on a shape to select it and then move it backwards or forwards by clicking on the up or down arrows.

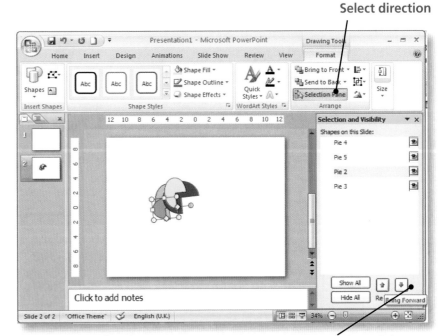

Fig. 5.45 Reorder shapes

Select direction

Click to move selected shape

5 To set transparency, select a shape and then click on the **Shape Fill** command and select **More Fill Colours**.

6 Set the transparency slider manually or enter a percentage – 100% will make it completely transparent.

Check your understanding 9

1 Start a new presentation that has two slides with blank slide layouts.

2 On slide 1, insert four circles and colour them black, white, light grey and dark grey.

3 Position them centrally in the four quarters of the slide by eye using the gridlines, guides and rulers to help you.

4 Now use the alignment options to position them exactly.

5 Group the shapes into one.

6 Copy the grouped shape to a second slide.

7 Colour the dark grey shape red.

8 Ungroup and then layer the shapes in the following order: black on top, then red, then grey, then white. Make sure part of each shape is available for selection.

9 Reorder the white shape so that it is one from the bottom.

10 Make the black shape 50% transparent.

11 Add an irregular shape such as a moon of any size to a separate part of the slide and colour it green.

12 Copy this shape and flip it to create a mirror image.

13 Position the two moons opposite one another aligned at the top of the slide.

14 Save the presentation as *Shape exercises*.

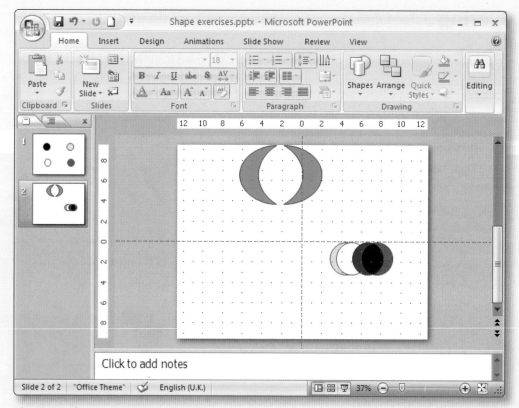

Fig. 5.46 Shape exercises

Tables

To add a table to a slide, you use the same process as when creating a table in a word-processed document. Once it appears, it can be resized, formatted and edited.

When the table appears, it will have a format set by default. On the **Table Tools – Design** tab in the **Table Styles** section you will find a range of different designs in the gallery that you can apply. You can also choose **No Style, Table Grid** for a plain table.

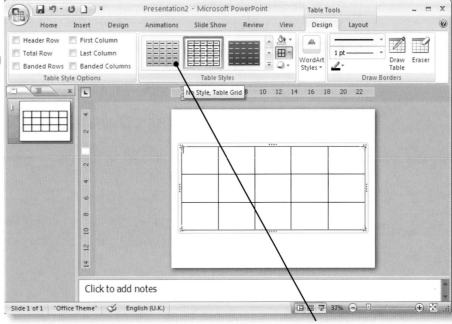

Fig. 5.47 Table style

Table styles

insert a table

1 Click on the **Table** command on the **Insert** tab.

2 Drag the mouse across the cells to select the correct number of rows and columns.

3 Let go of the mouse and the table will appear.

or

4 Click on the **Table** icon in the **Content** placeholder and enter the number of rows and columns in the boxes.

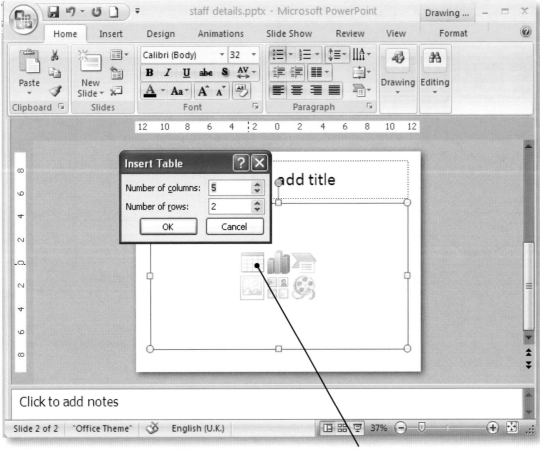

Fig. 5.48 Insert table using placeholder

Click icon

211

make overall changes

1 Move the table to a new position on the slide by dragging it with the mouse.

2 Resize the table by dragging out a border.

3 There is also a button on the **Table Tools – Layout** tab where you can resize the table exactly by typing in a new height and width.

Adding or removing columns and rows

If you need to make changes to a table you can do so in a number of ways.

add columns or rows

1 Click on a cell.

2 On the **Table Tools – Layout** tab, click on an **Insert** option such as **Above** or **Right**.

3 You can also click in the last cell and press the **Tab** key to add a new row.

remove columns or rows

1 Drag to select the unwanted cells.

2 Click on the correct option on the **Table Tools – Layout** tab after clicking on the **Delete** command.

3 Select the table border and press the **Delete** key on the keyboard to remove the table.

Click for options

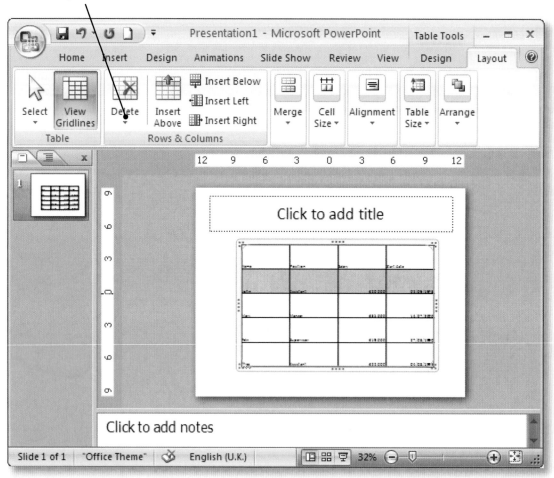

Fig. 5.49 Add or remove cells

Adding text

There are two ways to add text to a table:
- Type in each cell in turn.
- Copy data across from elsewhere.

copy in data from a data file

1 First create a table of the required size on a slide.
2 Open the data file containing the data and select all relevant cells.
3 Click on **Copy**.
4 Return to your slide and either click on a single cell or select all the table cells by dragging across with the mouse. They will appear light blue.
5 Click on **Paste**.

Formatting a table

As well as changing the appearance of data entered in the cells, you can also change the appearance of a table by adding different borders and shading, or removing borders altogether.

format cells

1 Select the cells containing the data.
2 Use normal formatting tools on the **Home** tab to apply different font types, sizes or to colour the text.
3 Use the alignment buttons to place the text horizontally in the cells. (See Unit 002 for details on using tabs in tables to align data within cells.)
4 For a deep cell, you can click on the **Align Text** button to set text vertically in the centre, top or bottom.
5 Drag cell boundaries to increase or decrease measurements, or enter exact measures in the cell size boxes on the **Layout** tab.

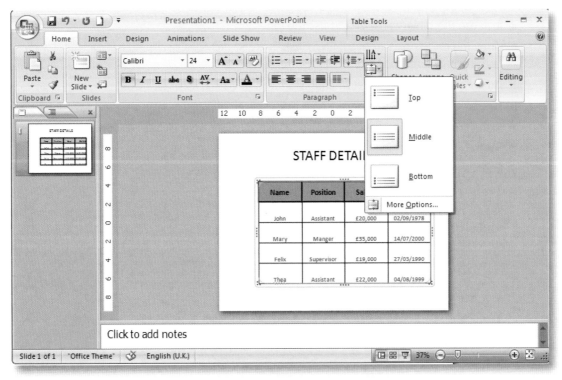

Fig. 5.50 Align text in cell

213

6 On the **Table Tools – Design** tab, set a type of border from the border button.

7 To remove borders, choose the **No Border** option.

8 Use the line style and width buttons in the **Draw Borders** group to change the type of line, making sure you click on the border button to apply the selection.

9 Add background colours to the cells from the **Shading** button.

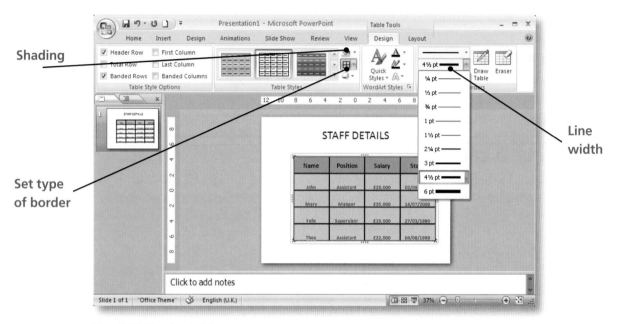

Fig. 5.51 Border table

Check your understanding 10

1 Start a new presentation.

2 On slide 1, create a table that has four columns and five rows.

3 Increase the table size so that it fills the slide.

4 Copy the data across from the Excel file *Staff List* provided on the CD-ROM accompanying this book.

5 Add the slide title *Recent staff changes*.

6 Format the column headings to Arial, font size 24.

7 Reduce the depth of these cells to about 1.5cm and centre align the entries.

8 Format all other data to Arial, font size 18.

9 Format the borders to a solid line with a width of 2¼ pt.

10 Shade only the cells containing the names green. All other cells should have no shading.

11 Save the file as *Staff details*.

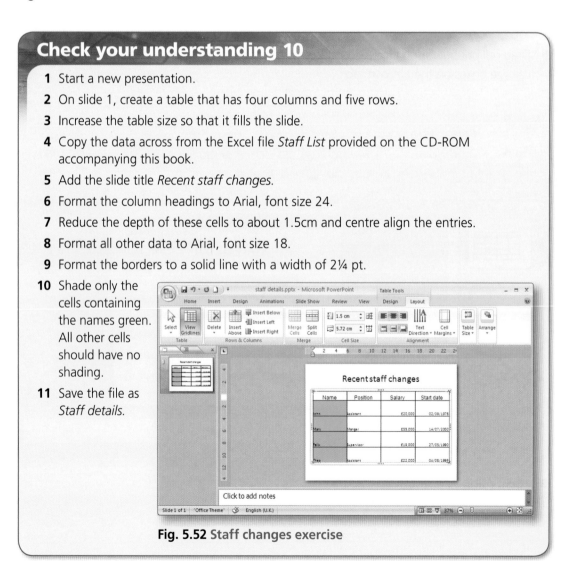

Fig. 5.52 Staff changes exercise

Charts

You will often want to display data on a slide in the form of a chart, and you can do this either by creating the chart directly or by copying across a chart from an application such as Excel.

create a chart on a slide

1 Apply a content slide layout and click on the **Chart** icon.

or

2 Click on the **Insert** tab and select **Chart**.

3 Select the appropriate chart type in the gallery that appears and click on **OK**.

4 You will be presented with a chart sample together with a spreadsheet containing temporary data.

5 Replace the data in the spreadsheet with your own and the chart will be created for you.

6 If you are offered too many columns or rows in the spreadsheet, select and delete them or they will confuse your final chart.

7 To amend the data at any time, reopen the spreadsheet by clicking on **Edit Data** on the **Design** tab or after right clicking.

Reopen spreadsheet

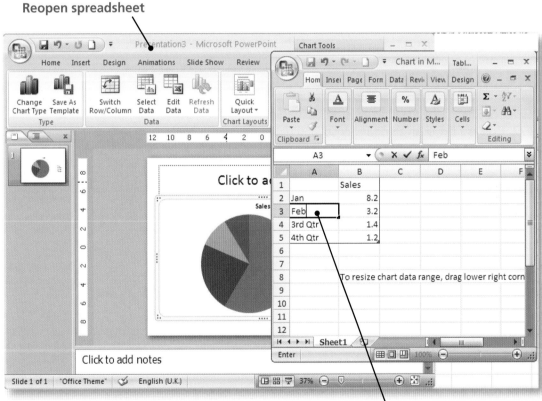

Fig. 5.53 Create chart

Enter your own data

Formatting the chart

The chart is created using Excel so you can use the familiar methods for making changes, such as:

- adding labels
- adding or removing the legend
- adding a chart title
- formatting colours of the data series
- changing chart type.

You can resize the chart on the slide by dragging out a border, or move it by dragging it with the mouse.

Add labels

Change chart

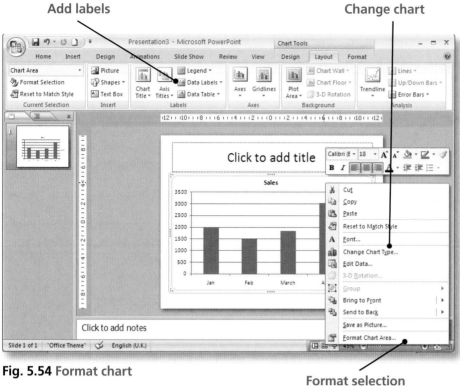

Fig. 5.54 Format chart

Format selection

format the chart

1 Click on the **Chart Tools – Layout** tab and add data labels or titles by clicking on the relevant buttons.

2 Right click on the chart or click on the command on the **Design** tab to select a different chart type.

3 Right click on a selected part of the chart or click on the **Format** tab to change colours and line styles.

Check your understanding 11

1 Start a new presentation and save the file as *Best sellers*.

2 Create a bar chart using the following data:

Event	Tickets
Christmas fair	350
Halloween party	56
Summer fete	426
Dog show	245

3 Remove the legend.

4 Change the title to *Ticket sales for best selling events*.

5 Increase the font size and make sure the title is well above the chart data.

6 Add X and Y axes titles: *Event* and *Ticket sales*.

7 Colour the data series red and the plot area yellow.

8 Save and close the file.

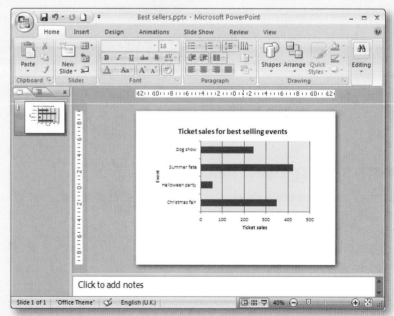

Fig. 5.55 Best sellers

216

Organisation charts/diagrams

A different type of chart you may want to add to a presentation is one that represents a hierarchy or family tree. Each person or item is displayed in its own box and these can have a particular relationship to other boxes in the chart.

insert an organisation chart

1 Click on the **Insert SmartArt** graphic placeholder.

or

2 Click on this option on the **Insert** tab.

3 Choose **Hierarchy** in the index and select your preferred layout. The first option is **Organisation Chart** and the second is **Hierarchy**.

4 When the chart appears, click in each box and enter your own text. The text will be resized to fit the box automatically.

5 Press **Enter** for entries on a new line in the same box.

Organisation chart

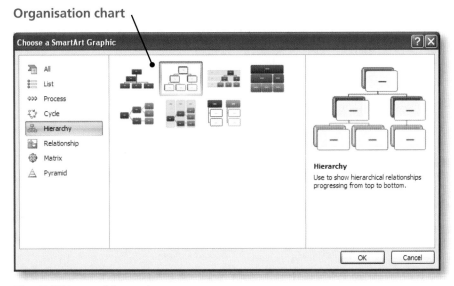

Fig. 5.56 Insert hierarchy

6 To add new relationships, click on a box and then click on the **Insert Shape** button on the **SmartArt Tools – Design** tab.

7 For the **Organisation Chart**, click on the drop-down arrow below the **Insert Shape** command to add people in roles such as Assistant (below but not directly below), Co-worker (at the same level) or Subordinate (directly below).

8 Promote or demote a new box if you want to move it to a new level by clicking on the appropriate arrow button in the **Create Graphic** group.

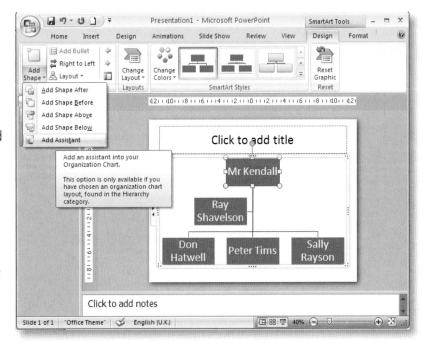

Fig. 5.57 Add assistant

9   You can also click on the **Text Pane** button to type in box entries directly.

10 Click on any box and press the **Delete** key to remove it from the chart.

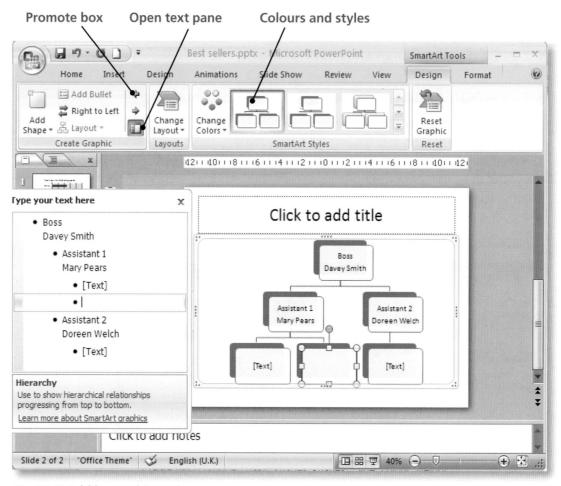

Fig. 5.58 Add box to hierarchy

Formatting the organisation chart

As the style of chart is set by default, you can select an alternative from the gallery or change individual boxes manually.

format the organisation chart

1   Set new colours or general styles from the gallery on the **Design** tab.

2   Click on an individual box to format or select several boxes by drawing round them with the arrow pointer.

3   Change shading, borders or shadow effects from the **Format** tab.

4   Select the text in any box (or several boxes at the same time) and use tools on the **Home** tab to format the text.

5   Click on and drag the box boundaries to make them larger or smaller, and click on and drag the outer blue border to change the size of the entire chart.

Check your understanding 12

1 Start a new presentation and save it as *Department*.

2 Create a slide showing the following organisation chart:

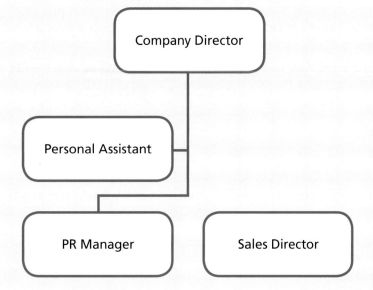

3 Now add a new Manager at the same level as *PR Manager* and *Sales Director* and add these details: *Head of Finance*.

4 Add a new linked box at an Assistant level: *Assistant to Head of Finance*.

5 Give the slide the title *Company Structure*.

6 Finally, remove the *PR Manager* role from the chart.

7 Save and close the file.

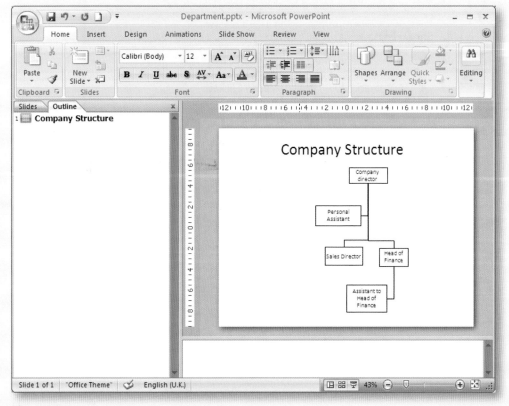

Fig. 5.59 Department

Adding sounds and moving images

To enhance your presentation, you may like to add appropriate sound effects or animated pictures. They can wake up an audience and are excellent when trying to entertain, but you need to handle them with care as they can distract or even irritate an audience if overused. (Note that as well as adding movies, the term **animation** applies to presentations where objects such as text boxes, bulleted lists, titles or charts are animated and build up piece by piece as the slide is displayed.)

Sounds are either saved WAV files or are chosen from those stored with the Clip Art or Media files. When added to a slide they are displayed as a loudspeaker. They only sound when you run a slide show.

add sounds

1 On the **Insert** tab, click on **Sounds**.

2 Select a sound file you have saved or open the **Clip Organiser**.

3 If you use the placeholder in a **Content slide** layout, it will take you to your computer files.

4 You can also go to **Insert – Clip Art** and make sure you select **Sounds** in the **Results should be:** box.

5 Enter keywords or scroll through the available sounds. These include bells, telephones, clapping and songs.

6 Click on any sound to add it to the slide. It will appear as a loudspeaker symbol.

7 Choose whether to hear the sound automatically or only when you click on the icon on the slide.

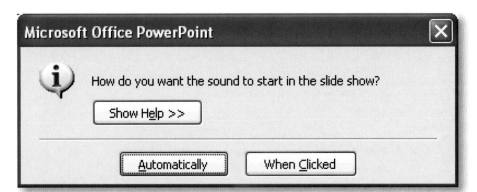

Fig. 5.60 How to hear sound

8 Click on the **Slide Show** button to hear the sound.

9 If you have chosen the automatic option, click on **Escape** to cancel the sound and close the slide show or **Page Down** to move to the next slide.

Loudspeaker symbol   Select sounds

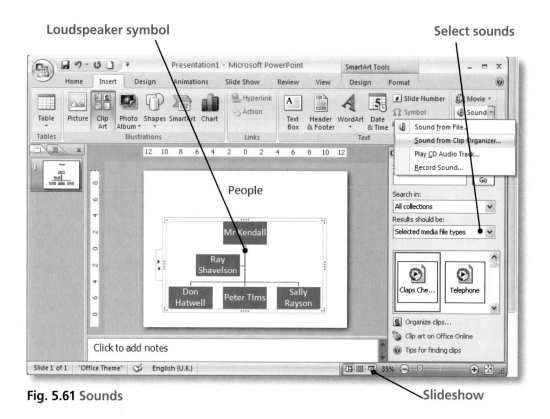

Fig. 5.61 Sounds   Slideshow

Fig. 5.62 Effect Options

10 To remove the sound, select and delete the loudspeaker symbol.

11 To hear the sound all through a presentation, click on **Custom Animation**. Click on the down arrow for the sound in the **Custom Animation** pane and select **Effect Options**.

12 You can now start the sound on the first slide and select on which slide it will end.

Where sound will end

Fig. 5.63 Play Sound

Insert media file

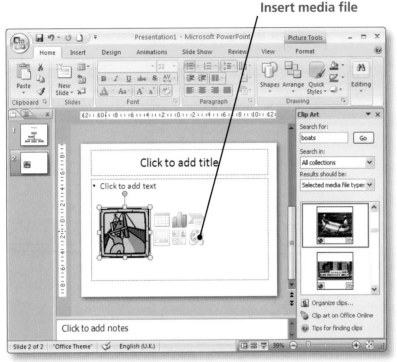

Fig. 5.64 Boat animation

Movies are normally added as moving GIF files. As with sounds, a range of animations are stored in the **Clip Organiser** or you can insert your own saved files.

add movies

1 Open the **Clip Organiser** and make sure you search for **Movies**, or click on this option on the **Insert** tab and look for animations on your computer or within **Clip Art**.

2 Click on an image to add it to your slide. It will look like a normal picture and can be moved and resized.

3 When you run the slide show, the picture will move.

Check your understanding 13

1 Start a new presentation saved as *Christmas.*

2 Give the first slide the title *Happy Christmas.*

3 Add a new slide with the title *Enjoy!*

4 Insert a suitable movie from the **Clip Organiser** and resize it so that it fills half the slide.

5 Insert a relevant sound from the **Clip Organiser** to start automatically when the slide appears.

6 Run the slide show.

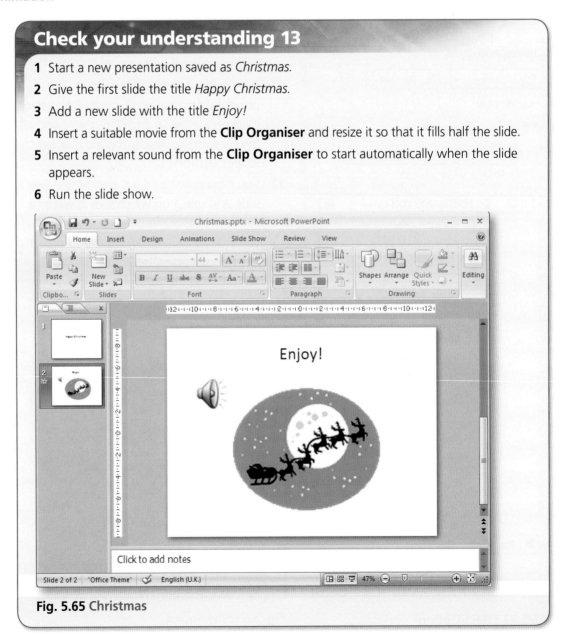

Fig. 5.65 Christmas

Animation schemes

Having decided on the main content of your slides, you can set your objects to appear or disappear as the presentation is run. A quick way is to apply animations to different types of object on the master slide so that they are translated throughout the presentation, or you can work on each slide to select the order in which each object will appear and decide how it will behave.

set animations

1 Open the slide you want to animate.

2 Click on the **Animation** tab.

3 Click on an object such as a chart, bulleted list or title.

4 Select an option in the **Animate** box such as **Fly In** or **Fade**. Rest your mouse on different choices to preview the effect or click on the **Preview** button.

5 You can set a time for the next slide to appear in the **Advanced Slide** section.

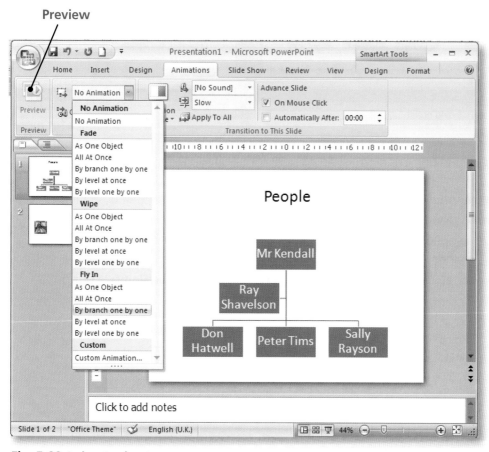

Fig. 5.66 Animate chart

6 For more detailed choices and multiple effects, click on **Custom Animation**. This opens an extra pane where you can choose:

- the type of effect to apply for its entrance or emphasis – for example, spin, grow or shrink
- whether it will appear automatically after or with a previous animation, or only on a mouse click
- its size and speed if relevant
- in what order each object will appear.

7 Click on the next object and set animation details. Numbers will appear showing the order in which each object is animated. Click on the **Re-order** button to change this if necessary.

8 Click on **Play** to preview, or run the slide show to see the effect.

Order   Set to appear automatically?

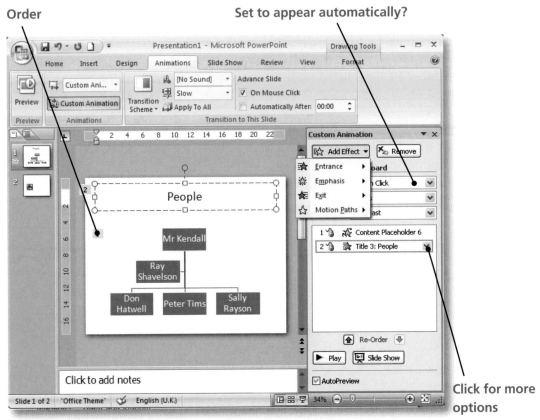

Click for more options

Fig. 5.67 Custom animation

9 To change some of the effects, click on the drop-down arrow for that object in the **Custom Animation** pane.

10 Click on an option such as **Timing** to open a new dialog box where you can make detailed changes.

11 Remove an animation by selecting the object and clicking on the **Remove** button or choose the **Remove** option from the drop-down arrow.

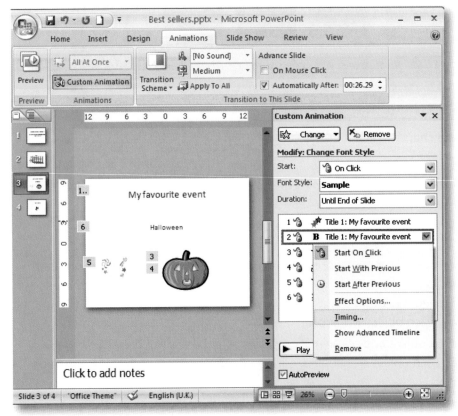

Fig. 5.68 Edit timing

Check your understanding 14

1 Open the presentation *Birds* provided on the CD-ROM accompanying this book.

2 Apply any animation to the bird image on slide 2.

3 Now apply a different animation to the slide title.

4 Run the slide show.

5 Change the order so that the title is animated first.

6 Save as *Bird animation*.

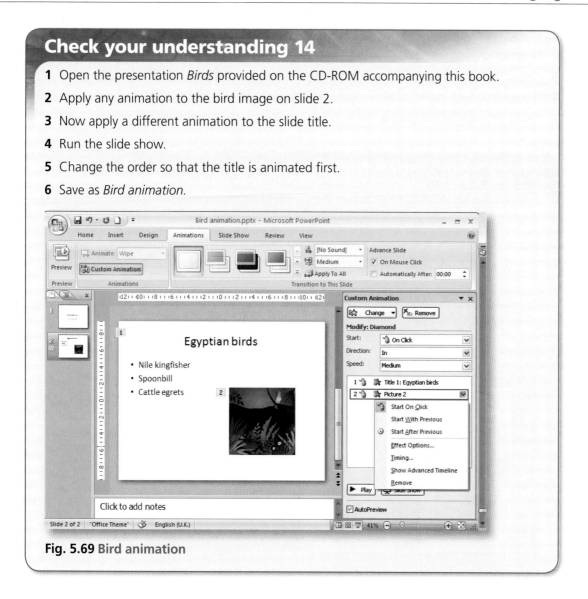

Fig. 5.69 Bird animation

Setting transitions

For a long presentation, you may like to automate the appearance of each slide or decide on a different way in which it arrives on screen. The change from one slide to the next is known as a transition.

set transitions

1 Click on the first slide and then click on the **Animation** tab.

2 In the **Transitions to This Slide** group, select a style of transition from the gallery. Arrows show the direction in which some effects will occur.

3 To remove a transition, click on the top **No Transition** option.

4 Set whether the transition will appear fast or slow.

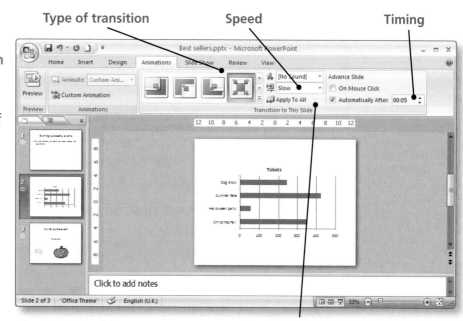

Fig. 5.70 Transition

5 If you don't want to click for each slide, deselect the mouse click option and set the number of seconds to wait before the slide will appear automatically in the **Timing** box.

6 Apply the same transition to all the slides or choose a different effect for each one.

7 If you view your slides in **Slide sorter view**, you will see that slides with animations or transitions applied, display a star and the timings of any transitions.

Check your understanding 15

1 Open the file *Walking* provided on the CD-ROM accompanying this book.

2 Set a different transition for each slide.

3 Make sure each slide appears automatically six seconds after the last one.

4 Save as *Walking slides* and close the file.

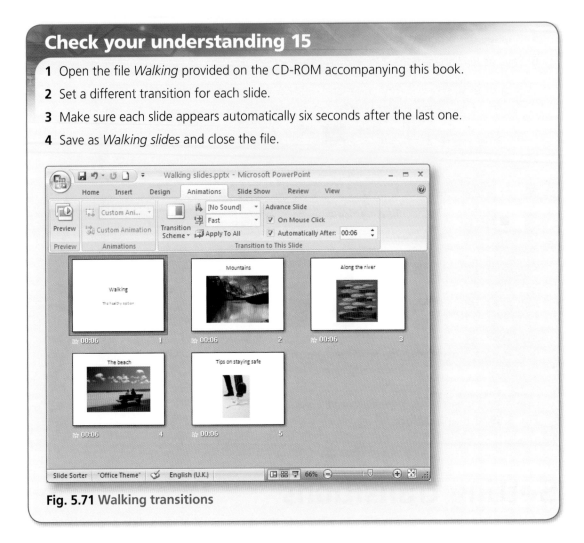

Fig. 5.71 Walking transitions

Professional slide shows

A professional presentation will not only contain slides with well set out points and suitable enhancements, but will also have an introductory slide and possibly summary (a slide listing all the titles in the presentation) and conclusion slides as well. You may also need to produce accompanying documentation such as handouts containing thumbnails of all the slides.

Once your slide show is ready, you can run it in different ways, e.g.:

- controlled by the speaker, so that each slide appears on a mouse click
- automatically, with each slide appearing after a set period of time
- unmanned, in a continuous loop – often convenient for a conference or exhibition
- selecting the order in which slides appear, rather than following the original order.

run a slide show in normal slide order

1 Click on the first slide and then click on the **Slide Show view** button, or go to **Slide Show – From Beginning**.

2 Either click to move to the next slide, or leave it to run based on timings set previously.

3 Make sure you have rehearsed the show so that timings are correct for the type of audience expected.

check timings

1 On the **Slide Show** tab, click on **Rehearse Timings**.
2 A timing box will appear in a corner of the screen and you can see how long each slide will be displayed and how long animations take to run.
3 Click on the arrow to move to the next object or slide, making sure you leave enough time for your audience to read any text or enjoy the effects.
4 At the end of the rehearsal, confirm the timings or click on **No** and make changes to your presentation.

Fig. 5.72 Rehearsal box

run a presentation in a loop

1 On the **Slide Show** tab, click on **Set Up Slide Show**.
2 In the dialog box, click in the **Loop continuously** checkbox under **Show** options.

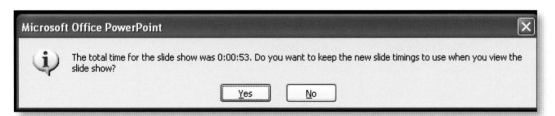

Fig. 5.73 Rehearse

3 Run the slide show and stop it running by pressing the **Esc** key.
4 Remove the tick in the checkbox to remove the loop.

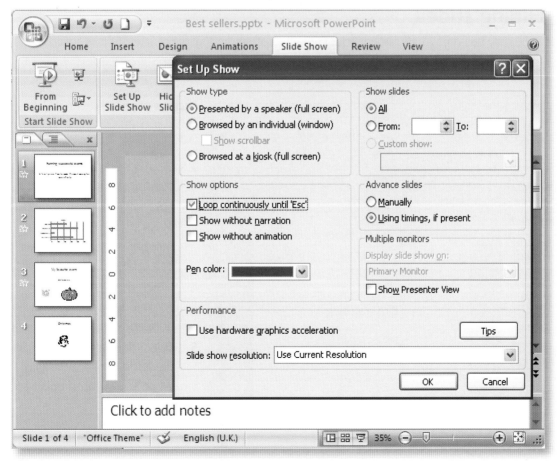

Fig. 5.74 Loop

change slide order during a slide show

1 Click on your preferred first slide before running the show, or start the show in the normal way.

2 At the point where you want to move to a different slide, right click the mouse on screen.

3 From the menu that appears, select **Next** or **Previous**, or click on **Go to Slide**.

4 All the slides in the show will be listed and you can click on the slide you want to display.

Fig. 5.75 Go to Slide

Hiding slides

There may be times when certain slides contain information that is not relevant to a particular audience. This slide can be hidden from view when the presentation is run, but still be available at another time.

Hidden slide

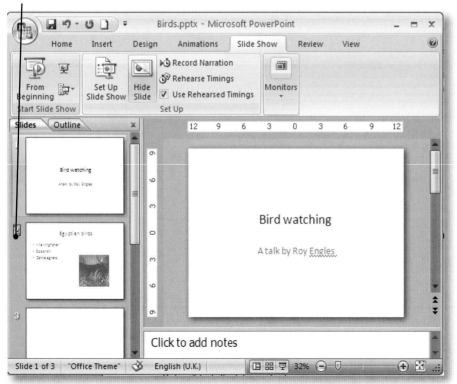

Fig. 5.76 Hide slide

hide a slide

1 Select the slide you want to hide.

2 On the **Slide Show** tab, click on **Hide Slide**.

3 On the **Slide** pane, or in **Slide sorter view**, its slide number will now be shown in a box.

4 When the slide show is run, hidden slides will be skipped.

5 Select the slide and click on the **Hide** command again when you want to unhide it.

Using hyperlinks

A different way to control a slide show is to embed hyperlinks in one particular slide that
will act as links to different slides (or even web pages or other files). In the same way that
hyperlinks work in web pages, if a slide hyperlink is clicked, it will take you to the linked slide.

You can either use text or images already present as the hyperlink, or add an action button.

create hyperlinks using slide content

1   Display the slide to contain a link and select the text or object to use as the hyperlink.
2   On the **Insert** tab in the **Links** group, click on **Hyperlink**.
3   In the **Link to:** pane, click on **Place in This Document**.
4   A list of slides will appear, so click on the slide you want to link to.
5   Click on **OK**.
6   When you run the show, hyperlink text will be coloured differently.
7   Click on the hyperlink to take you to the linked slide.

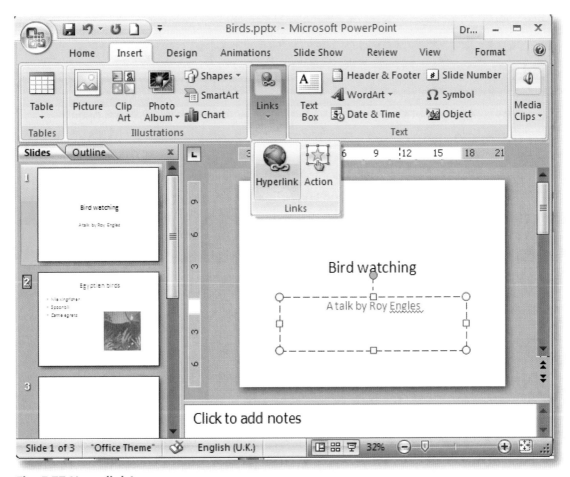

Fig. 5.77 Hyperlink1

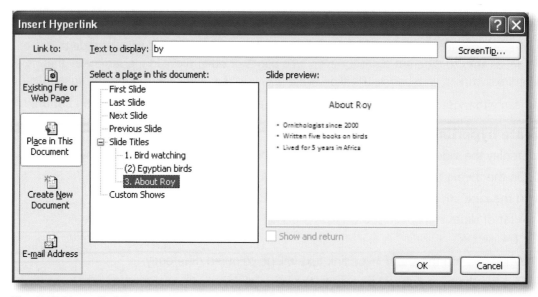

Fig. 5.78 Hyperlink2

add action buttons

1 Go to **Insert – Shapes**.

2 Select an **Action** button and add it to the slide. The hyperlink window will open automatically.

or

3 Insert and format any shape you want as the button and then go to **Insert – Action**.

4 When the dialog box opens, click on **Hyperlink to:** and select which slide to display when the link is made. This could be the next or previous slide or you can click on **Slide...** and select a slide number.

5 You can click or hover to activate the hyperlink, and can choose whether the shape should be highlighted when the mouse moves to it.

6 Click on **OK**.

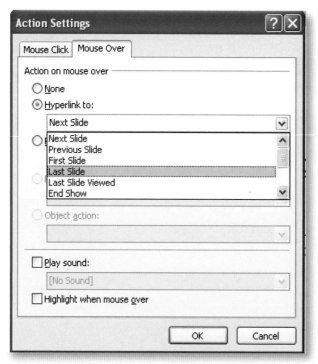

Fig. 5.79 Action

Check your understanding 16

1 Open the file *Letters* provided on the CD-ROM accompanying this book.

2 Animate the image on slide 1.

3 Add a link from slide 2 to slide 5 showing a chart, using the text *range of magazines* as the link text.

4 Run the presentation as a looped presentation.

5 Remove the loop and now start a normal slide show. Use the hyperlink to move from slide 2 to slide 5 before returning to slide 2 again to continue in the normal way.

6 Add transitions and make each slide appear after a five-second period.

7 Finally, hide slide 2 and check that it doesn't appear when you run the slide show.

8 Save as *Letters amended*.

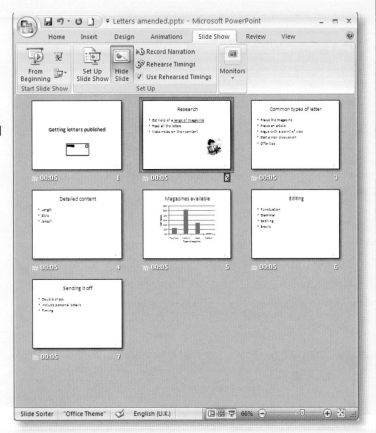

Fig. 5.80 Letters amended

Saving a presentation for use elsewhere

If you want to run a slide show on a computer that does not have PowerPoint 2007, it is possible to save it with software known as a viewer that will still enable it to be run. When you copy your Microsoft Office PowerPoint 2007 presentation to a CD, a network, or a local disk drive on your computer, Microsoft Office PowerPoint Viewer 2007 and any linked files (such as movies or sounds) are copied as well. (Previously you had to make use of a facility known as 'pack and go'.)

You can also save your file in a form that will open directly into a slide show, rather than Normal view.

save a presentation to a CD that will include the viewer

1 Click on the **Office** button.

2 Select **Publish**.

3 Click on **Package for CD**.

4 Click on **OK** when a warning appears as some files may need to be updated to run in the viewer.

5 Now name the CD and copy the files across.

Fig. 5.81 Package warning

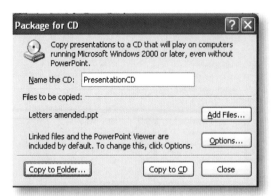

Fig. 5.82 Copy to CD

save a presentation as a show

1 Click on the **Office** button.

2 Select **Save As** and select the **PowerPoint Show** option.

3 Name and save the file in the normal way.

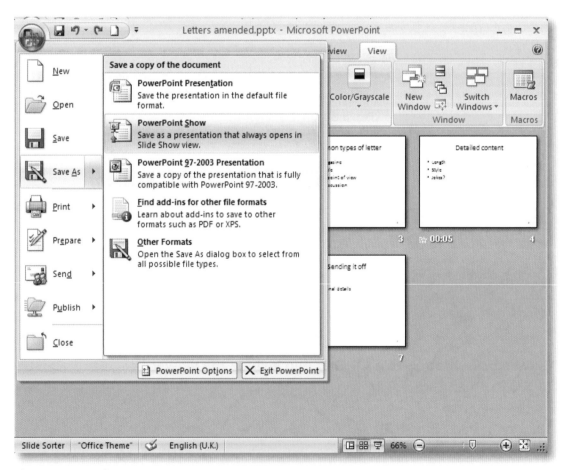

Fig. 5.83 Save show

Printing a presentation

You can print a number of objects from a PowerPoint file, including:

- all or selected slides
- handouts to accompany a talk
- speaker's notes
- the outline of the presentation.

Depending on what you want to print, you need to make sure the slide or page is set up correctly. Layout changes can be made when viewing the presentation in **Print Preview** or from commands on the **Design** tab.

create speaker's notes

1 Select any slide for which to make notes and, in **Normal view**, type your notes in the **Notes** pane.

or

2 Go to **View – Notes Page**. This view will show you how the page will appear when printed.

3 Type your notes in the text box presented below the thumbnail image of the slide.

4 If you need more room for the text or want the slide image larger, just drag the borders of the relevant box.

5 Repeat for any other slides where notes are required and then save the presentation.

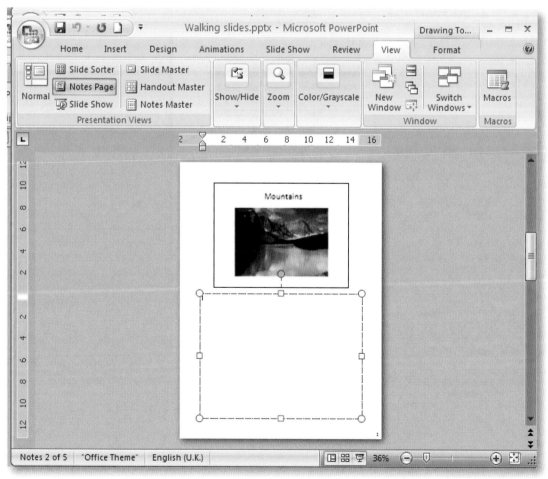

Fig. 5.84 Notes

make changes from Print Preview

1 Click on the **Office** button.

2 Select **Print – Print Preview**.

3 Click on **Options** if you want to print hidden slides, add or amend headers or footers or add a frame/border round each thumbnail.

4 You can also click on the **Colour/Greyscale** button to view how a coloured presentation will appear if printed in black and white.

5 Click on **Orientation** to change from Portrait to Landscape.

6 Click on **Zoom** to zoom in or out to check slide details or the overall effect.

7 Change the content of the **Print What:** box to select slides, handouts and so on or set how many thumbnails to a page.

8 Move through the presentation by clicking on the **Next Page** command.

9 Print the presentation by clicking on the **Print** command.

10 Return to **Normal view** by clicking on the **Close** button.

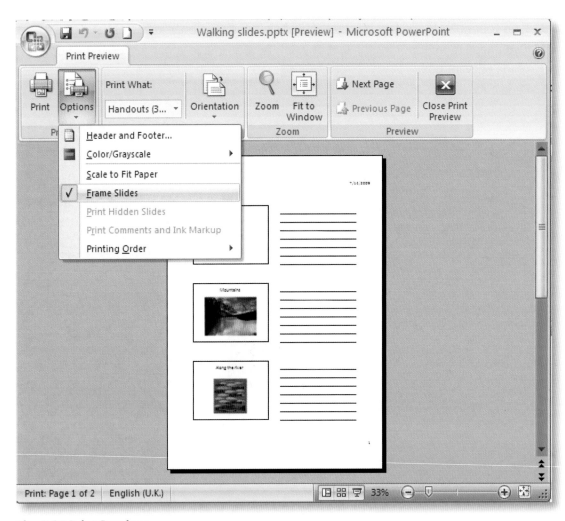

Fig. 5.85 Print Preview

change layout from the ribbon

1 Click on the **Design** tab.
2 Click on the **Slide Orientation** command to change slide orientation from Landscape to Portrait.
3 Click on **Page Setup** to open the dialog box. You can now select the type of printout you want and its orientation and size.

Fig. 5.86 Page Setup

print a presentation

1 Click on the **Office** button.
2 Go to **Print – Quick Print** to print a copy of all the slides, one per page, using default settings.
3 Click on **Print** to open the **Print** dialog box.
4 Here, select the type of printout, colour or black and white, the arrangement of thumbnails or number of copies you want.
5 To print only selected slides, enter slide numbers into the **Print range:** box or select the slides first – for example, in **Slide sorter view** by holding down **Ctrl** as you click on each in turn. Then click on the **Selection** option.
6 Click on **Preferences** to change your printer's settings.
7 If necessary, click in the checkbox to print hidden slides.
8 Click on **OK** to print.

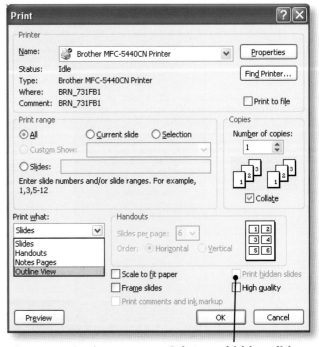

Fig. 5.87 Print box

Print any hidden slides

Check your understanding 17

Check your understanding 18

1 Open the file *Letters*.

2 Print handouts showing all the slides – four slides per page.

3 Print a copy of slides 1 and 3 only.

4 Finally, add the following note for slide 2: *Mention library opening times*. Print a copy of this page only.

5 Save the file as *Letter notes*.

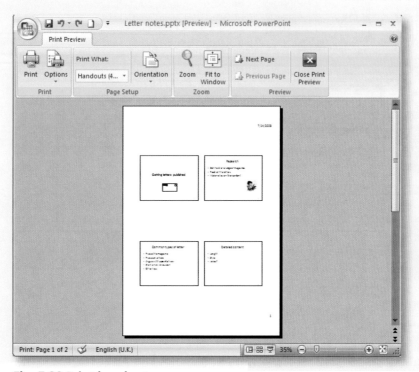

Fig. 5.88 Print handouts

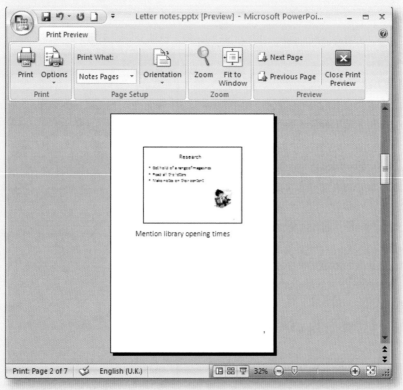

Fig. 5.89 Print notes

CLAiT Assignment

Task 1

1 You are going to create a five-slide presentation.

2 Use the master slide to set the following characteristics for your presentation:

    a. Set the orientation to portrait.

    b. Set a pale blue background.

    c. Use a Sans Serif font for all slides, e.g. Tahoma.

    d. In the footer, insert the automatic date and your name using a small font.

    e. Text: heading – large, bold font; 1st level – bulleted, medium size font and italic; 2nd level – bulleted, smaller but legible font, underlined.

3 Save as **Houses** and use it for all slides in the assignment.

4 Open the text file **Estate agency** and copy the prepared text onto the slides as shown. Save the presentation as **Houses** and ensure that the formatting of the master slide is applied to all the slides.

5 On Slide 1 titled INTRODUCTION:

    a. Insert the image **house**.

    b. Position the image below and to the right of the subtitle.

    c. Resize the image to ensure it does not overlap any text, keeping it in proportion.

6 On Slide 2 titled FINDING THE RIGHT HOMES, change text levels as follows:

    a. Terrace – demote to 2nd level.

    b. Bungalow – promote to 1st level.

7 On the same slide, add **Studio** at the 2nd level after **Flat**.

8 Print a handout of all the slides showing 3 slides per page.

9 Save the presentation as **Buying1**.

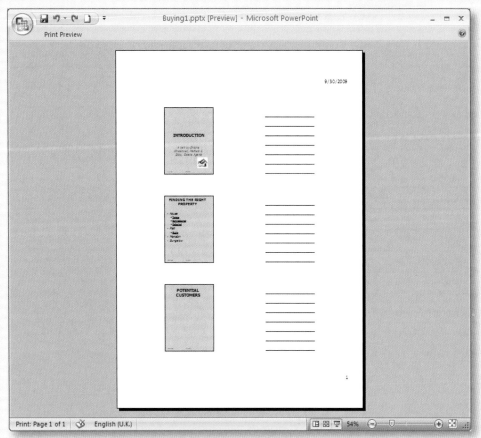

Buying1

Task 2

1 Insert a new slide as Slide 2.

2 Open the data file **Costs** and insert the information as a table with four rows and four columns.

3 Display all borders and make sure each entry fits on one line.

4 Format the text in a Sans Serif font such as Comic Sans.

5 Right align all column headings.

6 Format column headings in bold and italic and increase the font size so that it differs from the main text.

7 Give the slide the title: AVERAGE PURCHASING COSTS.

8 Save the presentation as **Buying2**.

9 Insert a new slide as Slide 4.

10 Use a layout to display an organisation chart. Add the slide heading **AGENCY STRUCTURE** and the following chart:

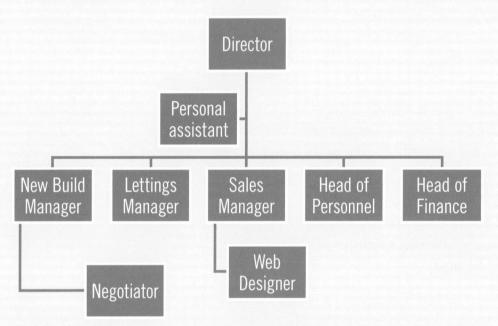

Organisation chart

11 Format all text entries in a Serif font such as Times New Roman in a legible font size.

12 Make sure all text is fully displayed with no words split.

13 Use a layout for Slide 5 that will enable you to include a chart.

14 Change the title to **ENQUIRIES** and use the following data to create a pie chart:

Type of Property	Numbers in 2009
Flat	5,230
Bungalow	3,365
Semi-detached	8,452
Detached	6,653

15 Give the chart the title: **Number of Homes Enquiries in 2009**.

16 Display data labels (categories) and percentages on the chart.

17 Remove any legend.

18 Format the text in a Serif font.

19 Carry out a spellcheck and proofread for any errors.

20 Throughout the presentation, change all instances of the word **Homes** with the word **Property**, maintaining case.

21 Save the presentation with the name **Buying3**.

22 Print a copy of the following slides only: Slide 2, Slide 3, Slide 5.

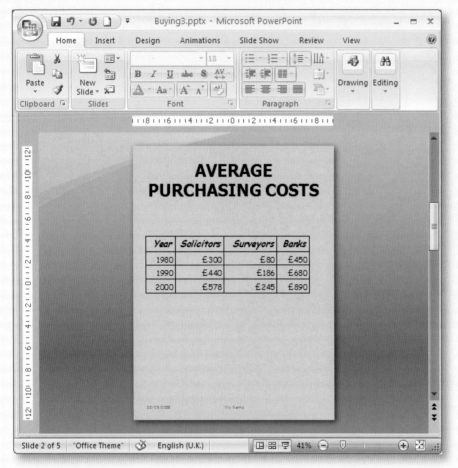

Slide 2

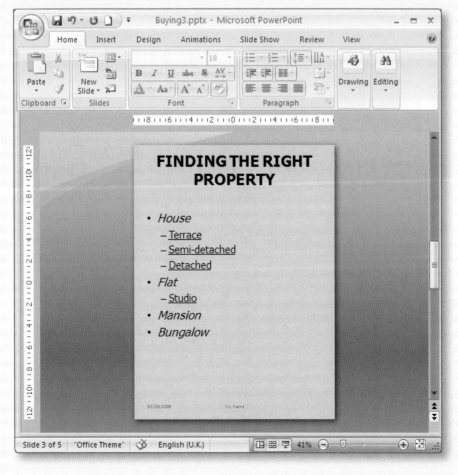

Slide 3

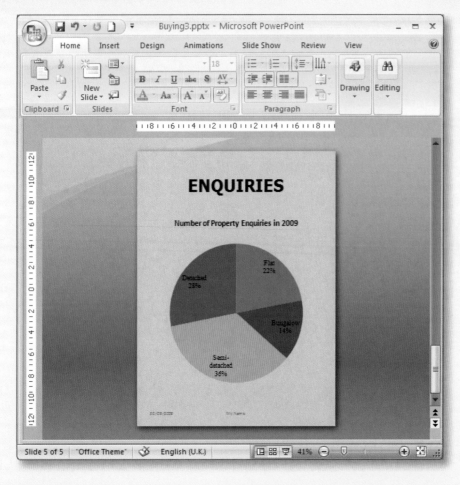

Slide 5

Task 3

1   Make the following amendments:

   a.   On Slide 3 (Finding property) delete the ***Mansion*** bulleted item.

   b.   Hide Slide 4 and take a screen print to show this has been done.

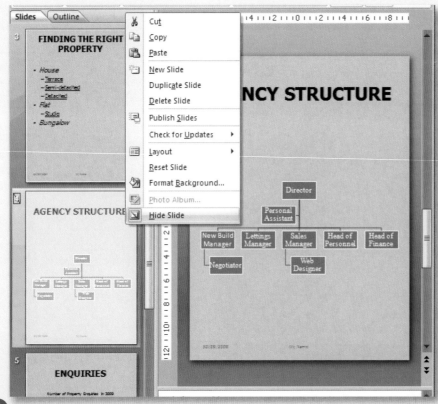

Hide slide

c. On Slide 3, create a hyperlink button to link it to Slide 5 and test that it works. Take a screen print to show you have created the link.

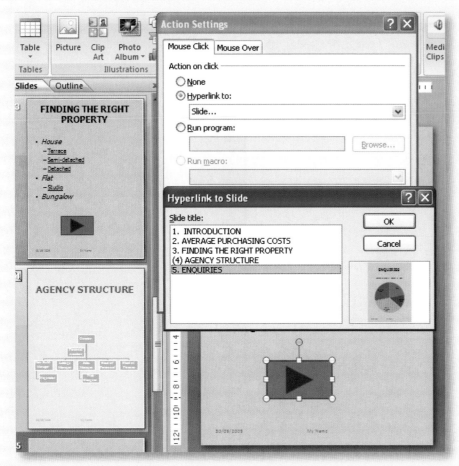

Hyperlink

d. Move Slide 3 so that it becomes Slide 2.

e. Animate the chart on Slide 5 and the image on Slide 1. Take a screen print in Slide Sorter view to show that these have been set, as well as the new slide order.

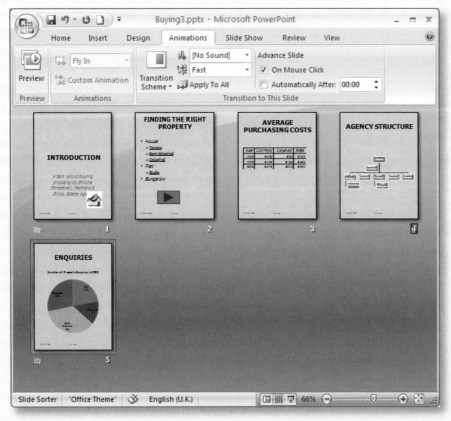

Animate

f.   Unhide Slide 4 and add a notes page, copying the text from the file **Staff**. Format the text in italic, green, medium font size and print a copy of speaker's notes for this slide only.

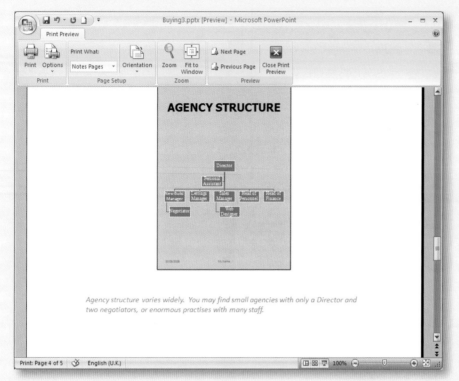

Notes page

2   Delete Slide 3.

3   Save the presentation as **Buying4**.

4   Close the presentation.

Task 4

1   Start a new three-slide presentation where there is a title slide and then two slides with a title and content layout.

2   Set the slides to Landscape orientation.

3   On the master slide, insert the image **handshake** and set it as the slide background to fill the entire slide. Make it semi-transparent.

4   On Slide 1 enter the text PREPARE FOR A JOB as the heading in a large font and copy this as the same heading to Slides 2 and 3.

5   Add the following subtitles in a medium font with no bullet points: (Slide 1) **Choose well**; (Slide 2) **CVs and covering letters**; (Slide 3) **Interviews**. Make sure the slides retain the same formatting and original layout.

6   Set timings and transitions as follows:

a.   Timings – 6 seconds.

b.   Transition – 1 per slide, all the same.

7   Take a screen print to show these settings.

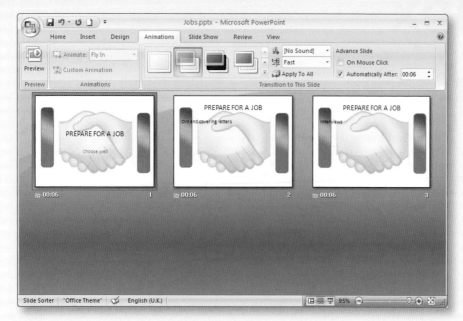

Transitions

8   Print a handout showing three slides on a page and where the background image is clearly visible.

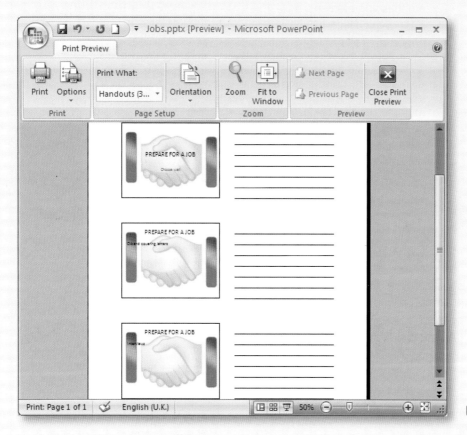

Handout with picture

9   Save the presentation as *Jobs* in a format that will mean it runs as soon as it is opened, and take a screen print to show this setting.

10   Save and close the file.

Index

merging cells 46–7
shade cells 51
using tabs 47–9
tabs
  columns with 39–41
  using menu 40–1
  using ruler 40
  Word tables 47–9
temperatures, typing 64
templates
  PowerPoint 186–7
  Word 3
text
  databases 133
  editing 61–70
(Excel) formatting 77, 80–1
insert symbols 63–4
move and copy 52
search and replace 61–2
setting in columns 39–50
special entries 63–4
superscript/subscript 64
Text Box
  charts 117–18
  PowerPoint 183, 191
Text Wrap 37
TIF files 35
time, adding 66
  see also date
transitions, PowerPoint 225–6
transparency, PowerPoint 188, 209

V

validation rule 145–7
vector images 35

W

widows and orphans 56
wildcards 158
WMF (Windows Metafile) 35

XY (scatter) graph 116–17

Yes/No box: database 133, 144

Z

zipped files 10–13